EXPOSED

EXPOSED

Environmental Politics and Pleasures
in Posthuman Times

Stacy Alaimo

 University of Minnesota Press
Minneapolis
London

Library of Congress Cataloging-in-Publication Data
Alaimo, Stacy, author.
Exposed : environmental politics and pleasures in posthuman times / Stacy Alaimo.
Minneapolis : University of Minnesota Press, [2016] | Includes bibliographical references and index.
Identifiers: LCCN 2016003043 | ISBN 978-0-8166-2195-8 (hc) | ISBN 978-0-8166-2838-4 (pb)
Subjects: LCSH: Human ecology—Philosophy | Human beings—Effect of environment on.
Classification: LCC GF41 .A3853 2016 | DDC 304.201—dc23
LC record available at http://lccn.loc.gov/2016003043

For Stephanie

Contents

◇◇◇

INTRODUCTION

Dwelling in the Dissolve

◇◇◇

The anthropocene is no time to set things straight. The recognition that human activity has altered the planet on the scale of a geological epoch muddles the commonsensical assumption that the world exists as a background for the human subject. New materialisms, insisting on the agency and significance of matter, maintain that even in the anthropocene, or, especially in the anthropocene, the substance of what was once called "nature," acts, interacts, and even intra-acts within, through, and around human bodies and practices. What can it mean to be human in this time when the human is something that has become sedimented in the geology of the planet? What forms of ethics and politics arise from the sense of being embedded in, exposed to, and even composed of the very stuff of a rapidly transforming material world? Can exposing human flesh while making space for multispecies liveliness disperse and displace human exceptionalism? What modes of protest, and what pleasures, do environmentalists, feminists, and other queer subjects improvise?

Exposed: Environmental Politics and Pleasures in Posthuman Times puts forth essays allied by their political commitments and their theoretical and methodological turns. It locates new materialist theories in fleeting ethical moments and particular political sites that make up the massive temporal and geographical expanse of the anthropocene. The book resists the temptation to engage in any sort of grand mapping or utterly lucid conceptualization, as that would be contrary to the embedded modes of epistemological, ethical, and political engagement that it traces. Yet the enmeshment of flesh with place runs through most of the chapters, suggesting a mode of being that deviates from the predominant Western mode of distancing the human from the material world. *Exposed* begins by considering the pleasures of inhabiting places where the domestic does not domesticate and the walls do not divide. It ends with an imaginary inhabitation of the dissolving shells of sea creatures that epitomize extinction in anthropocene seas. Dwelling in the dissolve, where fundamental boundaries have begun to come undone, unraveled by unknown futures, can be a form of ethical engagement that emanates from both feminist and environmentalist practices. Such practices are often improvisational, as activists, artists, and ordinary people seek to make sense of the networks of harm and responsibility that entangle even the most modest of actions, such as purchasing or disposing of any of the trillions of plastic objects circulating through the twenty-first century and thousands of years into the future. Paradoxically, while the temporal and geographic scale of the anthropocene is vast, the scale of human responses to environmental catastrophe is often minute. Rosi Braidotti describes a new ethical subject of "sustainable becoming," for example, who "practices a humble kind of hope, rooted in the ordinary micro-practices of everyday life: simple strategies to hold, sustain and map out thresholds of sustainable transformation."[1] While I am less hopeful than Braidotti, and more critical of the term "sustainability," as the conclusion will discuss, my own conception of trans-corporeality, which I extend in this book and which was set forth in my 2010 book *Bodily Natures: Science, Environment, and the Material Self*, does advocate the model of the ethical subject as one who is "rooted in the ordinary practices of everyday life."[2] *Exposed: Environmental Politics and Pleasures in Posthuman Times* locates subjects as they engage in both ordinary and extraordinary practices, both private, quotidian improvisations and more spectacular, even outrageous public performances. While *Bodily Natures* focused on environmental health and environmental justice movements, *Exposed* ranges across a motley mix of topics, including landscape art, performance art, naked protesting, marine conservation, plastic activism, and the scientific and popular encounters with "queer" animals. Even as one may not expect much humor in a book about environmental catastrophe, the art and activism analyzed here are often whimsical or jocular, perhaps because improvisation is playful, or at least cannot be authoritative or haughty. The Plastic Pollution coalition, for example, discussed in chapter 5, sponsors brief videos that are parodic and playful. La Tigresa, who strips for the trees in chapter 3, chases down loggers and shouts poetry at them through a bullhorn. And the ludicrous insistence by some scientists that various same-sex animals were not, in fact, doing anything sexual when they rubbed their genitals together, or the carbon-heavy masculinities that entail hanging metal "testicles" from trailer hitches, carry their own sort of humor. If we cannot laugh, we will not desire this revolution. But if there is a revolution happening, it is a modest one, often composed, as Braidotti puts it, of "the ordinary micro-practices of everyday life."³

The anthropocene is no time for transcendent, definitive mappings, transparent knowledge systems, or confident epistemologies. Surely all those things got us into this predicament to begin with, where presumed mastery over an externalized "nature" is all too triumphant, and yet also rebounds in unexpected, and usually unwanted ways. This does not mean I advocate phenomenological encounters or entirely local epistemologies. The immediacy of phenomenology, for example, does not enable trans-corporeal mappings of networks of risk, harm, culpability and responsibility within which ordinary Western citizens and consumers find themselves. While some of the activism and performance art that will be discussed stress the immediacy of the naked contact between body and place, these dramatizations are staged within a wider context of mediation and the horizons of scientific knowledges. Ursula K. Heise in *Sense of Place and Sense of Planet: The Environmental Imagination of the Global* argues, "Besides the valuation of physical experience and sensory perception, therefore, an eco-cosmopolitan approach should also value the abstract and highly mediated kinds of knowledge and experience that lend equal or greater support to a grasp of biospheric connectedness."⁴ Heise lucidly articulates this doubled reckoning with the local and the global, the immediate and the highly mediated, that is crucial for environmentalism. In her conclusion, Heise adds, crucially, that the "patterns of global connectivity, including those created by broadening risk scenarios," "are steadily increasing."⁵ Indeed.

Environmental crises demand scientific investigations, but what sorts of science is done, and how it is done, is deeply influenced by social, economic, and political forces, as science and technology studies have long insisted. This project is not a study of science per se, but it does analyze many works of science writing, and it tracks how activists, artists, and ordinary people engage with scientific data and perspectives. Popular science writing, whether it appears in books, magazines, websites, or social media, is one of the most crucial genres for environmentalism, yet it remains relatively neglected within the environmental humanities, ecocultural studies, and science studies. Chapter 5 contrasts Rachel Carson's environmentally oriented writing about the sea, which suggests a Darwinian community of descent, with Neil Shubin's *Your Inner Fish*. Although Shubin asserts the aquatic origin of the human, he nevertheless concludes with a utilitarian sense of the world as a vast apothecary for maintaining human exceptionalism. Popular science writing transmits not only facts and data, but also narratives, ideologies, values, ethics, politics, affect, and sometimes even a sense of species identity. Chapter 5 discusses how Captain Charles Moore juggles science, activism, and the need to captivate his audience as he dramatizes the devastating agencies of plastic, "man's surrogate," bits and nets of sinister substances that entangle us in responsibility. Chapter 6 analyzes the popular and theoretical reception of the proposed geological epoch, the anthropocene, against the iconic visual depictions of ocean acidification. Scientific conceptions, as they congeal into figurations, icons, aesthetics, or modes of seeing, circulate and reverberate across mainstream and activist media, shaping the terrain of environmentalism.

Living within "risk society," as Ulrich Beck argues, entails daily reckonings with scientific knowledge. And yet the citizen in risk society, Beck explains, suffers the "double shock" of not only hearing the news about, say, toxins in foods, but in the "loss of sovereignty over assessing the dangers, to which one is subjected."[6] I would like to recast this loss of sovereignty, a moment that erodes the sovereign individual subject, as an invitation to intersubjectivity or trans-subjectivity and even, though Beck did not argue this, to a posthumanist or counterhumanist sense of the self as opening out unto the larger material world and being penetrated by all sorts of substances and material agencies that may or may not be captured. Adriana Petryna's superb study *Life Exposed: Biological Citizens after Chernobyl* defines the biopolitical subject in terms of absolute exposure: "The deep intrusion of illness into personal lives fostered a type of violence that went beyond the line of what could be policed. There was no place that provided natural immunity from these unnatural and technical forces. Instead, there was a complete breakdown of immunities. This state of total unprotectedness constituted a baseline from which people in this world were refashioning themselves (and their bodies) as persons to be protected by the biopolitical regime in which they now lived."[7] While it would be mistaken to adopt the sense of exposure as a "state of total unprotectedness" for the scenarios, activism, and positions that will be analyzed in the chapters that follow, since they did not follow from events as catastrophic as that of Chernobyl, it does seem that an ethics and politics of exposure may be undertaken, by any informed and empathetic citizen, with precisely such catastrophes looming on the horizon. Exposures may be differential, uneven, or incommensurate; yet to practice exposure entails the intuitive sense or the philosophical conviction that the impermeable Western human subject is no longer tenable.

Performing exposure as an ethical and political act means to reckon with—rather than disavow—such horrific events and to grapple with the particular entanglements of vulnerability and complicity that radiate from disasters and their terribly disjunctive connection to everyday life in the industrialized world. To occupy exposure as insurgent vulnerability is to perform material rather than abstract alliances, and to inhabit a fraught sense of political agency that emerges from the perceived loss of boundaries and sovereignty. The loss of sovereignty that Beck discusses becomes not only epistemological but also onto-epistemological when it proceeds from a material feminist sense of the human as undeniably corporeal. This is the trans-corporeal subject I have advocated. Many of the essays in *Exposed* exhibit feminist occupations of a trans-corporeal subjectivity in which bodies extend into places and places deeply affect bodies. To dramatize oneself in place in this way is to critique the rational, disembodied Western subject's presumption of mastery or at least objectivity that is, supposedly, granted by detachment from the world. The exposed subject is always already penetrated by substances and forces that can never be properly accounted for—ethics and politics must proceed from there. And if penetration suggests something sexual here, all the better, as many of the essays trace protests and performances that involve exposure and other pleasures, occupying the scene of politics in ways that verge on the sexual and the queer. Those performances embody the crisis in rationality that feminism has uncovered again and again. But they also intimate that pleasure, desire, sensuality, and eroticism can pulse through the human exposed to place, permeating environmentalist ethics and politics as inspiration, catalyst, and energy. I resist the temptation here, however, to collapse these performances into one coherent queer or feminist environmentalism, and instead I let the trajectories of the separate chapters remain divergent, keeping my own desires for comprehensive theoretical mapping in check.

I have long believed that epistemological humility can function as a mode of environmental ethics that refuses utilitarian modes of mastery, so I welcome the recent turns to "unlearning" and even "failure" in academic circles.[8] In chapter 2 I argue that "queer" nonhuman animals elude modes of categorization, sparking an epistemological–ethical sense in which suddenly the world is not only more queer than one would have imagined, but also more surprisingly itself. The exuberant pleasures of thinking with, and feeling with, an abundantly, uncontainably queer world are countered by less enchanting ruminations. Reckoning with the anthropocene, climate change, the sixth great extinction,[9] ever-widening gaps between the rich and the poor within and between nations, neoliberal precarity, the resurgence of violent manifestations of racism in the United States, and, on a smaller scale, the threats to intellectual life and academic institutions means that there is so very much to unlearn. Eileen Joy writes, "Learning is always unlearning, a continual upending of everything you thought you knew, and therefore, difficult and melancholic," and yet within the present as a "creatively productive fugitive zone . . . we might practice the arts of divergent, tapestried becomings."[10] The "arts of divergent, tapestried becomings" beautifully describe the provisional practices of activists and artists analyzed in this book, as they engage with scientific data and schemas but resist the dominant impulse to externalize "the environment," and instead participate within the immediate, layered worlds they inhabit. Judith (Jack) Halberstam describes *The Queer Art of Failure* as a book about "alternative ways of knowing and being that are not unduly optimistic, but nor are they mired in nihilistic critical dead ends. It is a book about failing well, failing often, and learning, in the words of Samuel Beckett, how to fail better."[11] The modes of environmental activism that are analyzed in *Exposed* could be dismissed as failures, in that they hardly halt the carbon economy, the clear-cutting of forests, the devastation of ocean environments, or the proliferation of plastics. But within the scale of the anthropocene, surely all activism, all politics, all ethics, and all government policies will have been colossal failures. And yet, as Braidotti, insists, we nonetheless continue on, "for the hell of it and for the love of the world."[12]

Many of the ostensible "failures" this book investigates certainly "fail better," in that they are inventive, nuanced, impassioned, and intrepid. As a work of cultural studies, this analysis takes many popular and eccentric texts, artworks, films, and performances seriously, teasing out their complexities, and making sense of their embedded trajectories. One of the most quintessential works of cultural studies, in my view, is Laura Kipnis's analysis of *Hustler* magazine, where she calls Larry Flynt a "low-theoretician," and in which she asserts, at one point, that *Hustler*'s "intellectual work" is "on the order of the classic anthropological studies."[13] Halberstam describes "low theory" "as a mode of accessibility," but also as a "theoretical model that flies below the radar, that is assembled from eccentric texts and examples that refuse to confirm the hierarchies of knowing that maintain the *high* in high theory."[14] While *Exposed* emerges from several decades of engagement with high theory, the chapters often creep across the terrain of the low, featuring many modest and mundane sites, texts, and performances, and focusing on everyday encounters and practices. For the trans-corporeal subject, ethics and politics are always here and now, practiced through and within fraught, tangled materialities.

Elevated perspectives are problematic for both feminist and environmentalist visions, placing the human knower in a position above and beyond worldly entanglements.[15] Donna J. Haraway, in her now classic essay "Situated Knowledges," critiques the "conquering gaze from nowhere," the "view of infinite vision," the "god trick" of an unmarked, disembodied perspective.[16] Such a perspective has become all too commonplace in the predominant visual depictions of the anthropocene, as I will argue in chapter 6. I have found it quite fruitful to return to this early essay by Haraway, as twenty-first-century environmental, economic, and geopolitical panic has amplified the faith in floating perspectives—disembodied systems that can objectively map and maintain "resources" for some abstract global human subject of the present and future. Haraway contends that "feminist objectivity is about limited location and situated knowledge, not about transcendence and splitting of subject and object. In this way we might become answerable for what we learn how to see."[17] Many of the chapters that follow critique transcendence and the splitting of the subject and the object, countering this stance with alternative formulations of new materialist exposure. My conception of trans-corporeality was no doubt influenced by Haraway's feminist epistemology, as trans-corporeality originates with a recognition of the self as solidly located and denies the splitting of subject and object: the subject, the knower, is never separate from the world that she seeks to know. But that conception of positionality was deeply materialized through the process of editing (with Susan Hekman) *Material Feminisms* and writing *Bodily*

Natures.[18] Drawing on Karen Barad's theory, I developed a conception of the trans-corporeal subject who is "situated" in a more material manner, as the very substances of the world cross through her, provoking an onto-epistemology that reckons, in its most quintessential moments, with self as the very stuff of the emergent material world. Since transcendent epistemologies have fueled environmental destruction and harm to wild, domesticated, and laboratory animals, attention to the relation between epistemology and ethics remains vital. Barad's formulation of "ethico-onto-epistem-ology" calls us to consider the "intertwining of ethics, knowing, and being."[19] Indeed, her theory of agential realism makes epistemology an ontological matter. And, like many of the artists and activists analyzed in *Exposed*, Barad insists that humans "are part of the world-body space in its dynamic structuration," and that the "becoming of the world is a deeply ethical matter."[20]

Barad, the preeminent theorist of material feminism and the new materialisms more broadly, has given us the most thorough and robust account of material agencies as well as an ethics of mattering. Her thought will appear throughout this text, as I have found it dazzling and invaluable. As a work of cultural studies, however, this volume also delights in many activist texts that dramatize material agencies in unlikely ways. The Plastic Pollution Coalition artists and activists in chapter 5, for example, portray bits of plastic going rogue, rambling across the landscape, invading human and nonhuman bodies. It is significant that new materialist theories are developing at the same moment as many environmental activists seek to make sense of the interactions between consumer products, pollution, toxins, nonhuman animals, and humans. Jeffrey Jerome Cohen posits a relation between theories of material agency and activism: "Agency is distributed among multifarious relations and not necessarily knowable in advance: actions that unfold along the grid surprise and then confound. This *agentism* is a form of activism: only in admitting that the inhuman is not ours to control, possesses desires and even will, can we apprehend the environment disanthropocentrically, in a teetering mode that renders human centrality a problem rather than a starting point."[21] Many chapters that follow investigate how activists, artists, and others perform "disanthropocentrically," often by extending the human outward into particular locales or even by imaginatively dissolving the human as such. Even the structures of human habitats can be designed to be less anthropocentric as they open out toward the surprises of nonhuman agencies.

Environmental politics in the twenty-first century not only circulates through digital formats but permeates the material dimensions of everyday life. In their introduction to *Political Matters: Technoscience, Democracy, and Public Life*, Bruce Braun and Sarah J. Whatmore ask

> What if we took the "stuff" of politics seriously, not
> as a shorthand phrase for political activity but to signal
> instead the constitutive nature of material processes and
> entities in social and political life, the way that things
> of every imaginable kind—material objects, informed
> materials, bodies, machines, even media ecologies—help
> to constitute the common worlds that we share and
> the dense fabric of relations with others in and through
> which we live? What happens to politics—indeed to
> the "political" as a category—if we begin to take this
> *stuff* seriously?[22]

New modes of environmental activism as well as the provisional practices of ordinary, environmentally oriented people are engaged with these very questions, taking all manner of "stuff" seriously. Arguing for an "anarchic, ecologically informed ethics and politics," Mick Smith states that "ethics and politics as such are practiced every day, even in the most adverse of circumstances, often without ever having been formulated as such, and without seeking or requiring the permission of some higher authority."[23] I agree with much of Mick Smith's anarchic ethics and politics, which, through its attention to wildness, underscores the unpredictable vitality of nature as itself an anarchic force. But I think the affirmation of wildness needs to be complemented by a new materialist sense of the interacting or intra-acting material agencies of the objects, substances, and environments, produced by or at least altered by humans. Taking such "stuff" seriously mixes up the domain of ethics (primarily personal) and the domain of politics (primarily public), leaving us with something not unlike the feminist contention that the personal is political. A material feminist or new materialist environmentalism, however, would stress that the material interchanges between bodies, consumer objects, and substances become the site for ethical–political engagements and interventions.

Ethics and politics flow into each other, as the empty imaginary space for rational political debate becomes full to overflowing with all sorts of weirdly quotidian things that one would not expect to be there—plastic bags, cell phones, pesticides, bicycles, mercury-laden tuna. The public sphere needs to be reckoned with as if it were a landfill. If, as Barad contends, "the becoming of the world is a deeply ethical matter,"[24] that formulation could not be any more political than it already is, for inquiry into the nature of what is good must proceed to ask what is the world becoming and for whom? If the domains of the ethical and the political, the personal and the public, the domestic and the global, have collapsed into each other, they also reach across the unthinkable scale of the anthropocene as climate change, ocean acidification, extinction, and the production of xenobiotic chemicals make the location of each person's ethics and politics extend through vast geographical and temporal expanses, affecting countless species. The naked protestors of chapter 3 suggest as much; they dramatize the inseparability of human corporeality and the material world within particular locations, while hoping their performances will reverberate with political effects. Cohen notes, in *Stone: An Ecology of the Inhuman*, that the "ecological project of thinking beyond anthropocentricity requires enlarged temporal and geographical scales," yet "expanded frames risk emphasizing separations at the expense of material intimacies."[25] Many of the artists, writers, theorists, and activists that appear throughout *Exposed* endeavor to stretch material intimacies across immense scales.

For critical posthumanists and animal people, separating ethics and politics often makes little sense, as neither domain, within Western thought, has allowed space for concern over nonhuman lives. The relation between ethics and politics is a question that comes with such long histories that it cannot be resolved here. Nonetheless, it may be useful to point out that posthumanist new materialism, transcorporeality, and some modes of environmental activism muddle the categories of the ethical and the political, not only because they insist that nonhuman life is a matter of concern but because they demonstrate that even the smallest, most personal ethical practices in the domestic sphere are inextricably tied to any number of massive political and economic predicaments, such as global capitalism, labor and class injustice, climate injustice, neoliberalism, neocolonialism, industrial agriculture, factory farming, pollution, climate change, and extinction. The ethical and the political, like many other questions of and in the anthropocene, become matters of scale-shifting—improvisational interventions in lives and worlds where there is no stable background and nothing can be set straight. The mess we find ourselves in is perhaps most beautifully articulated by Stephanie LeMenager in her analysis of petrocultures, where even media—vehicles for the experience of "liveness"—are utterly reliant on oil: "We experience ourselves, as moderns and most especially as modern Americans, every day in oil, living within oil, breathing it and registering it with our senses. The relationship is, without question, ultra-deep."[26] Such saturated life worlds call for immersive practices and methodologies rather than dry, detached assertions.

Throughout this introduction I have kept "feminism" and "environmentalism" separate categories rather than collapsing them into environmental feminism or ecofeminism.[27] Environmental discourses are often drenched with the legacies of the term "nature." Since the concept of "nature" has long been enlisted to support racism, sexism, colonialism, homophobia, and essentialisms, it remains a rather volatile term, which feminists should approach with caution. Chapters 3 and 4, for example, analyze the vexed racializations of the exposed body within naked protest movements as well as the more overt evocation of blackness within an anti-environmentalist subculture. In the concluding chapter, Eli Clare notes the painful clash between living with an (incurable) disability and the discourses of ecological "restoration" projects. Furthermore, since most feminism and most queer theory is not environmentalist or oriented toward multispecies perspectives or concerns, there is no natural alliance here—and to echo Stuart Hall, no guarantees. And yet the theories, perspectives, texts, artworks, practices, and modes of being fabricated by those who have not been deemed as properly human do have something invaluable to contribute to posthumanism, inhumanism, the nonhuman turn, new materialism, critical animal studies, science studies, reflections on the anthropocene, and the environmental humanities. As these mushrooming areas are disciplined one hopes that the authoritative version of those fields does not marginalize the scholarship by and about those who have not been recognized as central to the (Western, Humanist) human. That would be terribly ironic. Moreover, the literature by indigenous and African American writers such as Louise Erdrich, Linda Hogan, and Octavia Butler has been invaluable for thinking human/nonhuman relations—I find myself returning to their work again and again. And the innovative work of such scholars as Sylvia Wynter and Kim TallBear will no doubt trouble and transform posthumanism and the material turn.

While I resist a position that unites environmentalism and feminism, preferring instead to be alert to the tensions, contradictions, and alliances—both within and between the two (political, subcultural, and academic) movements—I believe that their interrelations are generative—far beyond the territory where they overlap.[28] *Exposed* traces intersecting, allied, but also conflicting sites for feminism and environmentalism, presenting unsettling questions rather than comforting answers. Is La Tigresa a feminist when she offers up her bare breasts to male loggers to stop them from cutting down trees? Are feminist NGOs that address women's vulnerability in climate disasters environmentalist when they assume "nature" is a resource for domestic use? In *Undomesticated Ground: Recasting Nature as Feminist Space* I discuss feminist literature, theory, art, and activism that transform particular conceptions of "nature," in ways that are congruent with gender-minimizing and queer feminisms that destabilize the category of "woman." *Undomesticated Ground* argues for conceptions of nature that do not serve as foundations for gender essentialisms, racist taxonomies, or heteronormativity. The emergence of material feminisms, which retain the incisive cultural and political critiques of poststructuralist and postmodern feminism while making space for the active, emergent significance of the materiality of bodies, substances, and environments, has created new possibilities for productive alliances between environmentalisms and feminisms. Such alliances may negate essentialisms or dismiss them as crude or nonsensical.[29] Furthermore, as I argue in the conclusion, feminist theory, art, and activism, which has long contended with the paradox of proceeding from subjects who have been positioned as objects, is poised to make particularly potent posthumanist and new materialist interventions, given its history of thinking as the very stuff of the world. Material feminisms are at the heart of new materialisms and posthumanisms—not an offshoot, addition, or optional digression.

Throughout the book I argue that a material sense of exposure and pleasure fosters ontologies, epistemologies, ethics, and politics that interconnect the human with the nonhuman, the inhuman and the more than human. As a cultural studies project, *Exposed* takes activist and other "low" practices seriously, as inventive modes of political contestation. As a work within the environmental humanities, it grapples with climate change, biodiversity, sustainability, ocean conservation, environmental activism, and the depiction of the anthropocene. And as a study in new materialism it focuses on how the materiality of human bodies provoke modes of posthumanist pleasure, environmental protest, and a sense of immersion within the strange agencies that constitute the world. *Exposed* is organized into three sections. The first section, titled "Posthuman Pleasures," brings together two rather different essays. "This Is about Pleasure: An Ethics of Inhabiting" advocates pleasurable modes of environmentally oriented habitation, by way of landscape art, trans-species art, architecture, poetry, and science fiction. "Eluding Capture: The Science, Culture, and Pleasure of 'Queer' Animals" investigates why "deviant" pleasures of nonhuman animals are dismissed by both scientists and cultural theorists. The second section, titled "Insurgent Exposure," comprises two chapters, "The Naked Word: Spelling, Stripping, Lusting as Environmental Protest" and "Climate Systems, Carbon-Heavy Masculinity, and Feminist Exposure." This section analyzes the "carbon-heavy" masculinities that contribute to climate change as well as the forms of protesting that expose the human to the elements. While the masculinized, invulnerable body is promoted by both consumerism and U.S. nationalism, some forms of activism exhibit the naked body as a metonym of trans-corporeal connections between people and places. The third section, "Strange Agencies in Anthropocene Seas," argues for the importance of new materialism for ocean conservation movements. "Oceanic Origins, Plastic Activism, and New Materialism at Sea" critiques scientific origin stories that conclude with an inviolable human, turning instead to activists, scientists, and theorists who trace the strange agencies of substances that cross through humans and ocean creatures. The chapter "Your Shell on Acid: Material Immersion, Anthropocene Dissolves" pushes the idea of exposure, or radical openness to one's environment, to the extreme in an imaginary psychodelic dissolve—a figuration of anthropocene seas and their scenes of extinction.

This book (originally titled *Protest and Pleasure*), driven by many of my lifelong passions and political commitments, drenched in my theoretical obsessions, is not—oddly—something I set out to write. It is something that happened to me while I was trying to get to the sea. In hindsight, this seems utterly appropriate, given the arguments about immersion and strange agencies as well as the critiques of capitalist individualism and humanist detachment that course through the chapters. Many people, different scholarly communities, particular events, and vexing objects provoked this collection, and I owe an enormous debt to all of them, many of whom will not be adequately acknowledged here. Writing this book, even more than others, has underscored the significance of political communities and academic friendships. I cannot imagine thinking, or living, without them.

PART I

Posthuman Pleasures

1
This Is about Pleasure
AN ETHICS OF INHABITING

◇◇

*Despoilation of land in its many many guises is the
custom of the country.*

—Joy Williams, *Ill Nature*

Gregory Caicco introduces the collection *Architecture, Ethics, and the
Personhood of Place* with an excerpt from the Diné Hogan Song, which
reads, in part:

> haiye ne yanai
> It is placed, it is placed, it is placed,
> It is placed, it is placed, it is placed,
> Now at the Rim of the Emergence Place, it is placed, it is placed.

The song embodied, he says, "an intonation of our intent: to investigate alternative modes of ethical place-making within, beneath, or
outside modernity, if not the Western project as a whole."[1] The song's
insistent refrain, "it is placed," anticipates the devastating denial of
place in the contemporary United States. The taking of the land for a
terra nullius, an empty earth, has underwritten the assault on Native
American people and nonhuman habitats, spawning a multitude of
placeless places—no places—that are hardly utopian. Although people
who are not Diné are not the intended audience for this blessing, the
repetition of the condition of being placed provokes broader questions about inhabitation and ethics. My own emplacement within
North Texas incites despair at the new manifest domesticity of urban
sprawl, environmental racism, air pollution, water pollution, fracking,
and the ever more speedy destruction of habitats.[2] I will focus here,
however, on the confines of the home. That is, I will consider how
domestic space has served as the defining container for the Western

"human," a bounded space, wrought by delusions of safety, fed by consumerism, and fueled by nationalist fantasies. Would it be possible to redesign the domestic with an ethics of inhabiting such that the domestic does not domesticate and the walls do not disconnect? An ethics of inhabiting revels in the pleasure of interconnection and the joy of the unexpected; it embraces the possibilities of becoming in relation to a radical otherness that has been known as "nature."

Homeland Security

Gaston Bachelard wrote that "the house protects the dreamer, the house allows one to dream in peace."[3] Bachelard's implicitly male dreamer finds refuge within a comforting, feminine realm. Despite this cozy formulation, domestic space, in and of itself—but also in its oppositional relationship to the public sphere—has, for at least a couple of centuries, been a rather problematic construction for Western women. "Feminists," though the term may be anachronistic, have taken several divergent approaches to the problem of domestic space—a space that has, for the most part, been constructed to contain (privileged) women. Nineteenth-century American women writers who wrote "domestic fictions," as Nina Baym termed them, imagined the values and ideals of the domestic as a source of moral uplift for the wider culture.[4] At the turn of the twentieth century, activists in the women's club movement tried to break out of their domestic enclosure by way of "municipal housekeeping"; they created analogies between domestic skills and public work that would allow them to sweep their way into the public sphere. Conversely, as I discuss in *Undomesticated Ground*, a long tradition of North American women writers, theorists, and activists from the early nineteenth century to the late twentieth century turned away from domestic space and domestic values, turning instead toward nature as an "undomesticated ground," a place of freedom from the restrictive gender norms of the household.[5] While I would not underestimate the significance of the many innovative ways feminists have recast the concept of nature in order to create more hospitable habitats for feminist subjects, it is also important to consider the possibilities for reimagining domestic space as a suitable habitat for environmentalists. For the home is not only where most of us actually live—few of us follow the footsteps of Mary Austin's "Walking Woman," who wanders through the desert casting off societal values[6]—the home persists as a potent ideological space that has excluded feminists, GLBTQ peoples,[7] and other undomesticated creatures.

Nonhuman creatures, ecosystems, and environments have, no doubt, been harmed by the unrelenting encroachment of human domestic space, which is purified, as such, by the elimination of all but a few nonhuman species deemed desirable. The verb "domesticate," when it refers to the taming of animals, signifies both care and control. Notwithstanding recent work in animal studies that stresses the agency of domesticated animals and even plants, such as Donna Haraway's notion of the companion species, epitomized by dogs, who did nothing less than alter the course of human evolution, and Michael Pollan's apples, tulips, and marijuana that seduce humans into doing their bidding,[8] for the overwhelming number of animals who exist solely within factory farms, feedlots, and laboratories, domestication has resulted in confinement and suffering. As Yi-Fu Tuan explains, "Domestication means domination: the two words have the same root sense of mastery over another being—of bringing it into one's house or domain."[9] While the domestication of household pets in Western cultures tends to familiarize them, enlisting them into subordinate positions within the "family" of the human, and bringing them into the home or at least the yard, domestic territories are designed to keep wild creatures at bay, to ensure the domain of the human. An alternative, and rarely used, definition of domesticate, "to live familiarly or at home (with)," suggests that it is possible to imagine human habitation as living with, rather than walling out, other creatures.

In critiquing centuries of Western thought in which nature has been defined as the opposite of the human (with women, people of color, indigenous people, and the disabled placed in uneasy relations to both categories), many forms of contemporary environmental ethics and environmental feminism, such as the work of Donna Haraway and Val Plumwood, stress the concepts of kinship, continuity, and interconnection with nonhuman nature. Haraway's figure of the cyborg blurs the boundaries between humans and animals, nature and culture, offering us "a way out of the maze of dualisms in which we have explained our bodies and our tools to ourselves." Val Plumwood argues that to combat the persistent nature/culture, body/mind dualisms of Western culture we must "reconceive of ourselves as more animal and embodied, more 'natural,' and . . . reconceive of nature as more mindlike than in Cartesian conceptions."[10] Yet the home—with its walls, floors, and ceilings—is a bounded space, existing to keep the outdoors, outdoors, defining the human as that which is protected within. In fact, architecture itself is part of that dubious group of achievements used to distinguish the human from the animal, along with toolmaking and language. As Jacob Bronowski puts it, "Man is a singular creature. He has a set of gifts which make him unique among the animals: So that, unlike them, he is not a figure in the landscape—he is a shaper of the landscape. In body and in mind he is the explorer of nature. The ubiquitous animal who did not find but has made his home in every continent."[11] The fact that animals, including insects, are also architects of some renown does not diminish the degree to which the home, both literally and figuratively, has been erected as the spatial definition of the human. And, in most cases, the urban or suburban yard or garden of Western modernity is merely an extension of the house itself—fenced, bordered, and "clean." The home, the yard, the apartment complex, the gated community are places of mastery and careful demarcation of property lines—spaces of order and control.[12] An aesthetic of rigidity rules; surprise is verboten. "McMansions" gobble up extra-large portions of the outside world; windows are "treated" to domesticate their liminality. As Alberto Pérez-Gómez and Louise Pelletier put it, "Ours is a world of artifacts that are no longer a *bridge* between our consciousness and the external realities which we have not created; our artifacts seem to have rather created a *wall*, impossible to escape, surrounding us with our own dreams of control, self-referentiality, and cyberspace."[13] Before cyberspace, environmental artists sought to crumble the walls of self-referentiality, employing natural materials and irony. Walter de Maria's installation, New York Earth Room, for example, smuggles an illicit substance through institutional walls, presenting the placid spectacle of a load of soil dwelling indoors.[14] The smooth, white walls of the museum starkly contrast to the dark mass of dirt. An otherwise pristine and symmetrical room filled with soil foregrounds the material that is usually expelled; thus this installation manifests—in substance and space, appropriately—Luce Irigaray's deconstructive arguments regarding earth and woman as that which has been essential yet disavowed.[15]

We warm to the idea that "the house protects the dreamer," yet we may question, From what need we be protected? Despite the physical and psychological necessity for the safety of domestic enclosure, this seemingly benign dream of protection has morphed into a national delusion, with the borders of the home serving as a microcosm for racial, class, and national borders. Nan Ellin in her introduction to the collection *The Architecture of Fear*, published in 1997, notes that the "impulse to privatize is epitomized by the growth of gated communities, residential developments with controlled entryways and a clear separation from other neighborhoods, usually by a secure fence." These features of built environments, however, such as "gating, policing, and other surveillance systems," accentuate "a more general sense of fear by increasing paranoia and distrust."[16] Edward J. Blakely and Mary Gail Snyder call the "gated, walled, private community" a "new form of discrimination," in which the "frightened middle class" of the United States responds to school desegregation by "forting up."[17] The paranoid desire to "fort up" was intensified and authorized after the terrorist attacks of September 11, 2001. After 9/11, the U.S. Department of Homeland Security instructed its citizens to purchase duct tape in order to seal out the fumes of potential bioterrorist attacks. Ironically, in May 2001 George Bush had declared he would allow more arsenic in U.S. drinking water. Notwithstanding the tragedy of 9/11, for most people in the United States terrorism is a less likely threat than the toxins delivered daily through air, water, food, clothing, and commonplace household items. How much duct tape would it take to make actual homes as impermeable as the national fantasy of home? How can we seal out threats that are already within?

Notable artworks and films strip bare the illusion of domestic safety. The artist Lisa Lewenz, for example, depicted Pennsylvania's famous Three Mile Island Nuclear Generating Station, which suffered a partial core meltdown in 1979, in a calendar she produced in 1984. The station is framed by a homey kitchen window, photographed from within the supposed refuge of domesticity, warning that what is outside is always already inside, especially in the case of radiation. The flatness of the photograph erases any sense of distance between Three Mile Island and the surrounding kitchen window and wall.[18] The monthly calendar signals a spectacular disjunction between domestic time and radiation time. Similarly, the director Todd Haynes's 1995 film *Safe*, set inside wealthy, white, hyper-clean homes—"in a world," as he puts it, "as 'safe' (protected) and 'immune' (insulated) as you can hope to find"—exposes the assumption that the "home is a refuge" to be a dangerous delusion.[19] Here, people with environmental illness—the canaries in the kitchen—demonstrate that the domestic sanctuary is riddled with poisons. In one scene, for example, the actress Julianne

Moore's character attends a baby shower. Sitting stiffly in the pristine, carpeted living room, she begins wheezing and bleeding from the nose, struggling to breathe but maintaining perfect posture, as the camera moves menacingly closer to her and the music threatens. The benign baby shower morphs into a scene from a horror film, a monster movie horrifically lacking a visible monster.[20] The film includes a microcosm of environmental justice principles, as the domestic workers do not enjoy the privilege of dwelling in the expensive home and yet are exposed more directly to the cleaning chemicals employed to keep it pristine.

As the manifestation of the patriarchal family, consumerism, and widening economic disparities, the house erects boundaries between nature and culture, wild and civilized, imperiled and supposedly "safe." Good housekeeping, tidy landscaping, and every manner of pest control shore up these borders, which distinguish middle- and upper-class order from the chaos that lurks outside. The normal sense of the domestic is deeply infused with what Simon Estok has called "ecophobia," as the domestic walls off bad nature, the "menacing alterity of the natural environment."[21] The home, however, is never impermeable: electricity, media, advertising, and consumer goods stream in along with water, air, particulate matter, human inhabitants, guests, microbes, and nonhuman pests. The same sorts of poisons sprayed on the lawn will make their way back into the house in the tap water. The home as a conceptual apparatus, however, where we dwell within supposedly impermeable walls, may undergird the sense that environmentalism is only about protecting distant places or calendar-pretty species, and not about the visceral recognition that our air, our water, our food, indeed our very cellular matter, circulate through the very same lifeworld as that of the plankton or whale. Is it possible, then, to reenvision the home—a place constructed of literal and metaphorical walls—as a liminal zone, an invitation for pleasurable interconnections? What would it mean to recognize the claim of nonhuman animals on human territories? What sorts of practices or pleasures would foster posthuman, anticonsumerist subjectivities?

Lively Inhabitations

Poets, science fiction writers, and filmmakers have all imagined alternative visions of the home. Rather than solid, dead walls that demarcate, they imagine walls as the stuff of life. In this more than zoomorphic or biomorphic but rather, simply, biological architecture,[22] the walls come to life as an interface, a zone of the intermingling of nature and culture. The sensual pleasures of encounter, contact, and intimacy may lure us away from the temptation to wall ourselves off within a narrowly human sphere, even as painful histories saturate embodied recognitions.

The poem "The History of Red," by Linda Hogan, saturates the color with American Indian history, juxtaposing birth and colonialism, enmeshing land and people, as babies are born wearing the "red, wet mask of birth," delivered onto a "land / already wounded / stolen and burned."[23] Jodi A. Byrd explains that for "American Indians, who have lived for tens of thousands of years on the lands that became the United States two hundred and thirty years ago, the land both remembers life and its loss and serves itself as a mnemonic device that triggers the ethics of relationality with the sacred geographies that constitute indigenous people's histories."[24] In Hogan's poem the temporal reach of geographical relations supplants diminutive U.S. histories with American Indian remembrances that are not only more vast but more immediately present. In the second stanza, the speaker asserts an earthy and material ancestral consciousness, of "human clay," which remembers "caves with red bison / painted with their own blood."[25] The penultimate stanza reflects on homes made from the skin of bison:

> Red is the human house
> I come back to at night
> swimming inside the cave of skin
> that remembers bison[26]

The human house acknowledges its indebtedness to the bison, as the animal's blood remains so palpable that stepping inside is akin to swimming. Despite the transformation from bison to human dwelling place, the animal's liveliness persists, since the cave of skin still remembers. For Hogan the domestic does not exclude history, blood, or animal vitality and remembrance. With American Indian experiences of genocide, war, colonization, dispossession, confinement, and forced removal from homelands, the question of inhabitation resonates, sharply, across centuries and continents, but also back to a time before colonization. Furthermore, the human house, dependent on an animal that was nearly rendered extinct by European Americans, also suggests the precarity of nonhuman lives that may only be remembered by their own remains. The Native American home does not expel nonhuman nature, as the bison is the literal material of the walls and the metonymic environment within. The skin of the animal is the skin of the home described as cave, a naturally occurring shelter for human and nonhuman life. These palpable metonymic intimacies extend geographically and temporally, and suggest, by sharp contrast, the historical, vanished violence of manifest destiny.

Hogan's poem "The Bricks" begins with a seemingly solid and inert building block and reveals the life within. Although the architect mentioned in the poem did not intend for the bricks to possess a life of their own, they harbor "desperate seeds" from the "edge of the forest."[27] The seeds become stowaways, inhabiting the bricks as if they were still the soil:

> There is a secret longing
> inside bricks that holds worlds
> together, a forest dreaming
> inside every wall,
> wanting to send out
> a passionate tendril of life
> as in Japan
> when humans emptied other humans
> of their lives.
> Cities fell
> and bricks flowered
> with plants from distant mountains[28]

The structure of this passage, in which the reference to the horrific nuclear bombing of Hiroshima stands as an exemplar within another sort of chronological frame, suggests the resilience of the forest.[29] Hogan wanders from the confines of the home, looking back to a time when "We did not yet / live inside walls of bricks."[30] She animates the life history of the bricks, beginning the poem by speaking of the "ancestors" of the bricks, who came from places of wild gourds, salamanders, and snakes. She grants the bricks a confidence born of abundance, as the clay "would stand upright / knowing a forest lived inside it."[31] The bricks are not discrete objects placed upon the land to build the house, but instead, they are of the land, of the forest. They not only "hold worlds together" but hold the potential to decolonize the land, the forest, the people, as they somehow flower despite violence and dispossession: "destruction would bloom."[32]

How human habitations and geographies are drenched with histories, memories, and nonhuman agencies has been a vital question for the speculative fiction author Octavia Butler as well. Katherine McKittrick argues that Butler's novel *Kindred* "allows us to imagine that black geographies, while certainly material and contextual, can be lived in unusual, unexpected ways." Being in place, does not, McKittrick argues, entail stasis, but instead, "being materially situated *in place* holds in it possibilities that do not neatly replicate or privilege traditional geographic patterns of geometry, progress, cartography, and conquest."[33] Butler, in her Xenogenesis series of science fiction novels (*Dawn, Adulthood Rites*, and *Imago*), depicts a complicated geography of postapocalyptic human settlements that have been overtaken by aliens with radically different ontologies of habitation. The alien Oankali she depicts live as part of a biological world in which the homes and even the spaceships are living, sentient creatures. This allows Butler to imagine modes of inhabitation that that are not premised on seeing the world as a warehouse of inert building materials, but instead, as multispecies cooperatives. The alien Oankali species do not build their homes or construct their spaceships; they grow them, or rather, they urge them to grow themselves. Butler eschews instrumental relations by creating a world in which the habitat is a creature that lives in a symbiotic relation with the beings it houses. The homes and the ships share memories and DNA with the beings that inhabit them. Jodah, a part-Oankali, part-human character in *Imago*, introduces his environment, which, significantly, has a name: "Lo was parent, sibling, home. It was the world I was born into . . . woven into its genetic structure and my own was the unmistakable Lo kin group signature."[34] Kinship is accompanied by the desire for difference since the Oankali take sensual pleasure in "tasting" different genetic structures (in contrast to the humans, whose hatred of difference makes them maim and murder). An ethic of inhabiting is suggested by the hopeful conclusion of the series, Jodah's "planting" of a new town from a single cell. Planting a new town, while more agricultural than architectural, however, still repeats histories of colonialism and conquest. And yet, since it is an alien species, the Oankali, that drive the multispecies spaceships and initiate the encounters between beings and worlds, it is hard to know whether this echo of colonialism is a sort of poetic justice for the dominating humans who are shown to be violent and xenophobic. To turn to the matter at hand, Butler's speculative abodes foreground and animate a physical world that is most often presumed to be the mere background or resource for the human. There may be colonialist echoes, ironic or not, but the world is not depicted as a *terra nullius*. Rather, Butler's speculative fictions provoke us to imagine ways of inhabiting that do not shore up the boundaries of the human by rendering other lives into dead material.[35] McKittrick's question resonates here: if "our expressive demands can demonstrate a new worldview, in what ways can ethical human geographies, or interhuman geographies, be mapped?"[36]

The low-budget film *Habitat* is also set in a postapocalyptic world; the ozone layer has been depleted to the extent that no biological life can exist outdoors.[37] The mad scientist Hank Simes transforms his typical suburban home into a biologically diverse habitat. His neighbor, by the way, just happens to be a "green architect." As the neighbor tells Hank, green architecture "fuses the living space with the environment." Hank responds, "How about doing it the other way around? Bring nature indoors?"[38] As Hank transforms his home into a lush jungle, thick with green vegetation, fungi, molds, and oozingly organic walls, his wife, Clarissa, happily wanders about, free from domestic labor. It would be impossible to clean such a place, nor need she shop or cook when nutritious fare can simply be plucked from the walls.

These three visions of the home envision it as a living place of surprising pleasures. Connection, interrelation, and intersubjectivity are the ontological conditions from which new delights and new ethics emerge. Pleasure spirals through these ethical ontologies that are unmistakably material rather than abstract, disembodied principles. For Hogan and Butler, memory itself is woven into the walls, history is corporeal, and sensuality becomes a practice and a praxis. These dwellings arise from the dream of an unmediated relation between human and nature; the walls do not contain, they bestow.

These poetic habitations may seem distinctly apolitical in that they magically eschew capitalism and consumerism. Yet they provide a crucial topos for thinking otherwise, for acting in other ways, perhaps even "*cultivating subjects . . .* who can desire and inhabit non-capitalist economic spaces."[39] Not only do they implicitly critique a consumerist culture that renders the natural world a dead resource but they may also, in their tangible immediacy and intimate scale, counter what Stephanie LeMenager has called "petrotopia." She describes "petroleum-utopia" as "the now ordinary U.S. landscape of highways, low-density suburbs, strip malls, fast food and gasoline service islands, and shopping centers ringed by parking lots or parking towers," which has been manufactured by the "relentless production of space."[40] LeMenager argues that petrotopia "represents itself as an ideal end-state, the service economy made flesh, repressing the violence it has performed."[41] Petrotopia, presumably, offers many pleasures, and environmentalism would be remiss were it to counter only with asceticism and self-denial. As Catriona Sandilands explains, "It is not only that abundant pleasure is virtually absent in (most) ecological discourse, but that it is often understood as downright opposed to ecological principles; frugality and simplicity appear to act as antithetical principles to enjoyment or generosity."[42] Thus, if capitalism conscripts every desire by clothing it within the uniform of consumerist demand, predominant environmental discourse tells us to resist those desires. The problem with this sort of ethical appeal, however, is that it is hardly appealing.

Dan Phillipon counters the "earnestness and joylessness" of sustainability discourse by drawing on Wendell Berry's notion of "extensive pleasure," in which aesthetics, pleasure, and politics come together in the practice of savoring food that is made more delectable with the "accurate consciousness of the lives and the world from which the food comes."[43] The notion of extensive pleasure parallels the pleasures of inhabiting that I am advocating here, in that even something as intimate as eating can be experienced as already dependent on and interwoven with (sustainable, organic) food production. Of course such pleasures are difficult to sustain given that many people do not have access to or cannot afford sustainably grown, organic foods and even the most ecofriendly agricultural practices, as well as nearly all human dwellings, for that matter, still take habitat from wild creatures. And yet, a wholesale swath of (justifiable) despair may be lifted by someone such as Ron Finley, the urban guerilla gardener from Los Angeles, who transforms neglected urban spaces such as parkways into vegetable gardens for the community.[44] Finley's gardens, as well as the broader DIY movements for urban gardening, urban bee- and

chicken-keeping, community gardens, community-supported agriculture, seed saving, wildscaping, seed bombing, biodiesel conversion,
and biking, are extensive sorts of nonconsumerist, often community-
oriented environmental pleasures that reduce one's carbon footprint
through abundant practices of sensuality and playful experimentation.
Whereas the subject within an ecological ethic of deprivation denies the body, its playfulness and desires, Gail Weiss advocates the
development of "a moral agency that can only be experienced and
enacted through bodily practices, practices that both implicate and
transform the bodies of others."[45] An ethics of inhabiting could, then,
begin with corporeal practices, practices that interconnect and transform. We would experience the "walls" of the body, of the human
self, as permeable places of connection, just as Butler, Hogan, and
the film *Habitat* imagine the walls of the home. Moira Gatens describes Spinoza's understanding of the human body: "It is a body that
is in constant interchange with its environment. The human body is
radically open to its surrounds and can be composed, recomposed
and decomposed by other bodies."[46] Forms of place-based activism,
especially environmental activism, may deliberately stage themselves
in ways that dramatize how human corporeality and human practices
are immersed within environments and affected by particular, embedded encounters. After a year and a half of living in a giant redwood
named Luna in order to save it from logging, Julia Butterfly Hill notes
the changes in her own physical being:

> The tree had become part of me, or I her. I had grown
> a thick new muscle on the outer sides of my feet from
> gripping as I climbed and wrapping them around
> branches. My hands had also become a lot more muscu-
> lar; their cracks from the weathering of my skin re-
> minded me of Luna's swirling patterns. My fingers were
> stained brown from the bark and green from the lichen.
> Bits of Luna had been ground underneath my finger-
> nails, while sap, with its embedded bits of bark and duff,
> speckled my arms and hands and feet. People even said I
> smelled sweet, like a redwood.[47]

Luna's patterns become imprinted on Butterfly's hands; traces of
Luna lodge within her own flesh. The practice of tree-sitting, which

has often captured media attention, can itself be seen as a sort of uto-pian mode of inhabiting, since the structures that the tree-sitters erect in the branches make a minimal impact on their environment. The redwoods dwarf these puny platforms, which are only inhabitable, surely, because they open out onto the forest. One tree-sitter, in the film *Tree Sit: The Art of Resistance*, gleefully relates, "We're in the mid-dle of this—it lives, it breathes!"[48] Corporeality becomes, then, inter-corporeality, as described by Gail Weiss: "To describe embodiment as intercorporeality is to emphasize that the experience of being em-bodied is never a private affair, but is always already mediated by our continual interactions with other human and nonhuman bodies."[49]

The tree-sitters, who risk their lives and bodily integrity climbing hundreds of feet up in the giant redwoods, surviving terrible storms and the constant threat of forcible eviction, are inspiring in terms of environmental politics, but a rather extreme example of an inter-corporeal mode of inhabiting. The philosopher Ladelle McWhorter provides a more down-to-earth model. She tells of how she took up gardening because she desired the taste of a real tomato. In order to procure that pleasure, she needed to feed the dirt that fed the tomato plant. She nearly fed the dirt some Dorito crumbs, but then she no-ticed their ingredients and decided she could not "feed that crap" to the dirt: "Dirt and flesh. Suddenly it occurred to me that, for all their differences, these two things I was looking at were cousins—not close cousins, but cousins, several deviations now removed."[50] Realizing that dirt is her cousin does not domesticate it. Instead, this realization de-familiarizes her own body, in that her flesh is suddenly understood as a relative of the dirt. Significantly, McWhorter's use of the words "deviation" and "deviant" throughout her book link her own sexual identification to the environmental ideal of biodiversity while queer-ing the generative process of evolution itself. McWhorter's newly dis-covered kinship with the dirt illustrates her Foucauldian ethics, which emphasizes the possibilities of becoming that emerge from cultivating pleasurable practices. She asks, "What pleasure-developing practices might I cultivate that will enable me to resist, oppose, and counter sexual regimes of power?"[51] Similarly, in an article titled "Desiring Nature, Queering Ethics," Catriona Sandilands argues that "if we take seriously the argument that the ecological crisis is, even in small part, a problem of desire—specifically, of its narrowing, regulation, erasure, ordering, atomization and homogenization—then, I think, queer theory has a great deal to offer environmental ethics, and vice versa. . . . Desire for, and pleasure in, the tactile presences of the Other has the potential to reorient sexuality away from both ecologically and sexually destructive relations."⁵²

We are not accustomed to thinking about pleasure as ethical— political, perhaps, but not ethical. Yet, if ascetic practices frequently enforce corporeal boundaries or encourage a denial of or transcen- dence from nature and place, pleasurable practices may open up the human self to forms of kinship and interconnection with nonhuman nature. Mary Oliver, in her poem "Wild Geese," charts an ethical path that abandons injunctions to be good. Instead, she advocates a sense of kinship with wild geese and other natural creatures. This sense of interconnections begins with human corporeality: "You only have to let the soft animal of your body / love what it loves."⁵³ Oliver's poem, with its anaphora, "You," begins by interpellating a new sort of ethi- cal subject, one who is not subjected to repentant rituals but instead responds to the call of the nonhuman world. Ethical action arises, then, from the recognition of one's specific location within a wider, more-than-human kinship network. In the final lines the "you" is no longer anaphoric, no longer capitalized; rather it is gently embedded within the voice of the world.

The foundation for pleasure, Oliver suggests, is to receive our "place in the family of things." What would this entail? If we begin with our first mode of inhabiting, that of inhabiting our own bodies and then experience those bodies as permeable, as open to surround- ing human and nonhuman bodies, we can conceive of a corporeal ethics: an ethics that is always "in place" and never a disembodied or free-floating Cartesian affair. This ethics-in-place counters the unsustainable romance of wilderness fantasies and the lure of eco- tourism, which may lead us to neglect the beauty and worth of the wildness that exists in the actual places we inhabit.⁵⁴ A compensatory movement needs to emphasize human dwellings as habitats, reveal- ing our interconnections with nonhuman nature and the possibilities for a multitude of sustainable pleasures. Conversely, we must recog- nize animal cultures, animal memories, animal pleasures, and animal homes, making space for them within all-too-human landscapes, as it is no longer possible, within the anthropocene, to imagine they will survive somewhere else.

Multispecies Art and Architecture

Is it possible, then, to reimagine domestic space as a ground for corporeal ethics and as a multispecies habitat? Dwelling in bodily pleasures may erode the foundations of humanism, as human exceptionalism resides in the exultation of the mind and spirit, removed from the flesh. As Rebecca Solnit argues, "In a very real sense we can be described as living inside the heads of others, in an excess of interiority that obliterates our own relation to material origins, to biologies, to our bodies."[55] Edward Casey, in *Getting Back into Place*, draws on Merleau-Ponty's concept of the "flesh of the world" to argue that "my body and natural things are not just conterminous but continuous with each other. . . . In such flesh the fibers of culture and nature compose one continuous fabric."[56] This image is a compelling one, especially since much of the scholarship on corporeality, including Weiss's book on embodiment as intercorporeality, quoted above, ultimately cuts the fabric in an all-too-human pattern, severing the threads that would weave us into nonhuman nature.

Scholarship in architecture has, historically and for the most part, restricted itself to the domain of the human.[57] There is no mention, for example, of green architecture or environmental ethics in Sherry Ahrentzen's comprehensive overview of feminism and architecture.[58] And even though Margaret Somerville asks, "What is architecture walling in and walling out in a material or tangible sense?" nonhuman animals and environments do not figure into her answer. More disturbingly, she concludes her essay with a spaceship metaphor that dreams of an ultimate escape from the earth: "Architects and architecture will play an important role in designing and building the tangible and intangible spaces—the 'spaceships'—in which we travel in body, mind, and spirit into the future."[59] Similarly, in a chapter titled "The Voices of Space," Karsten Harries forecloses the significance of the environment by asserting an entirely anthropocentric system of evaluation, claiming that "for architecture at least, the human being is the measure of all things."[60] Even most scholars who develop a corporeal ethic do not extend that ethic in such a way that it takes nonhuman nature into account. See, for example, the essays in Dodds and Tavernor's *Body and Building: Essays on the Changing Relation of Body and Architecture* and those in Bloomer and Moore's *Body, Memory, and Architecture*.[61] Deborah Fausch, in "The Knowledge of the Body and the Presence of History: Toward a Feminist Architecture," envisions an architecture that would "provide modes for the experience of the body, give validity to a sense of the self as bodily—a sense that may be shared by both sexes." She contends that the "discipline of architecture can thus provide room to address Western culture's tendency toward abstraction, distortion, mistreatment, even banishment of the body."[62] While I would endorse this vision, I would extend it in such a way that corporeality is recognized as interconnected with the natural world. Indeed, it is hard to imagine a better way to combat "Western culture's tendency toward abstraction" than by attending to the precise materializations of both corporeality and the workings of nonhuman nature. Furthermore, as I will argue in the conclusion, predominant paradigms of sustainability, including those within architecture, employ a managerial sense of distance, aiming to construct technical fixes to environmental problems. The human in this scenario is an expert, a problem solver, an engineer, a rational, calculating entity who is not vulnerable, fleshy, or interconnected with material processes, but stands outside, constructing the world. The surprises, the actions, the agencies of the material world may flip the gestalt here, disrupting how the environment—built, constructed, happened upon, or transformed—serves as the mere ground for the human figure. Elizabeth Grosz, in *Architecture from the Outside*, insists that nature is, in fact, the very "stuff" of architecture, in at least two ways: "Architecture relies on a double nature—nature as standing reserve, as material to be exploited and rewritten, but also a nature that is always the supersession and transformation of limits and thus beyond the passivity of the reserve or the resource, nature as becoming or evolution."[63]

It is this second sense of nature as "becoming or evolution," of course, that is usually ignored, repressed, or battled in a culture bent on excessive consumption and control. In order to loosen the grip of "the human" on the human, however, we must cast ourselves out into that sense of nature as becoming, unprotected by delusions of perfect enclosure. Just as an ethics of inhabiting entails turning the human outdoors, it also entails, conversely, inviting what is out of doors in. This reversal is not meant to level out everything into an anthropocentric or anthropomorphic sameness that denies nonhuman nature its own particular beings and agency but, instead, to make space for an "ethics of the Real," in Catriona Sandiland's term, in which nature is not a "human constructed presence, but an active, enigmatic other."[64]

Such an ethics would entail making space for animal memory and animal culture. In 1968, for example, Alan Sonfist proposed that "public art in urban centres throughout the world could include the history of their natural environment," by naming streets after plants, birds, and other animals, and having sounds of birds and other animals and even the smell of plants "emitted into the street."[65] If we extend Foucault's concept of "countermemory" to nonhumans, we can imagine the stories of those who lost, the creatures who no longer inhabit what are now city streets, announcing their claims. While the histories of plants and animals deserve public monuments, the risk, of course, is that many people would prefer piped-in bird songs—which would entertain without dropping feathers or feces—to live, unpredictable birds.[66] Similarly, while biomorphic or zoomorphic architecture may inspire environmental practices, creating a building in the shape of a bird does not substitute for creating a building in which birds, many of whom have been displaced from their habitats by human buildings, may comfortably nest.

Although it is crucial to remember nonhuman claims on the land, it is more important to create space for animal habitats and to practice rewilding. An environmental ethic of inhabiting would, first and foremost, respect the habitat requirements of nonhuman creatures. This would range from the common practice of creating bird, butterfly, lizard, frog, or mammalian habitats in the backyard to reenvisioning the city as a "zoöpolis." Jennifer Wolch argues that "the reintegration of people with animals and nature in zoöpolis can provide urban dwellers with the local, situated everyday knowledge of animal life required to grasp animal standpoints or ways of being in the world, interact with them accordingly in particular contexts, and motivate political action necessary to protect their autonomy as subjects and their life spaces."[67] In one sense zoöpolis demands an entirely different outlook from citizens, city planners, park administrators, architects, and landscape architects. In another sense, however, the concept may incite a recognition of the creatures that are always already in our midst. In my city neighborhood in Dallas, for example, raccoons and possums peak into the bedroom window and green anole lizards peer into the kitchen from their tree. Once, a tarantula, startled by the dog, ran into the living room. After looking around a bit, it calmly strolled out the door that I was holding open with a broomstick. News reports tell us wild turkeys have invaded New York City, bald eagles have been introduced into Manhattan, and park officials attempt to acclimate owls in Central Park. Coyotes are appearing in cities, even in upscale New York neighborhoods. Lance Richardson, refuting the common idea that wild animals such as coyotes "are an aberration, to be marveled at and then segregated off in . . . an 'appropriate wilderness area,'" argues that we should coexist with urban coyotes, which help control rodent and feral cat populations.[68] Mark Weckel, cofounder of the Gotham Coyote Project and conservation biologist, proposes that "in our particular period of time, coyotes could be more championed as a flagship species for urban environments."[69] Urban, suburban, and exurban environments must be reimagined as multi-species habitats rather than the exclusive territory of humans and their pets.

The 2001 documentary *Winged Migration*,[70] which follows several different species of birds as they migrate around the globe, shows the world from the birds' perspectives as they struggle to find food, water, and a place to rest within continents that are increasingly made up of inhospitable stretches of cities, suburbs, blacktop, industrial zones, and agricultural monocultures.[71] The anthropocene is a merciless world for migrating species. Seeing from the perspective of migrating birds, who cannot simply remain in their designated place but must, every year, undertake tremendous journeys, can underscore how the places we inhabit belong to other creatures as well. Agricultural, industrial, urban, suburban, and exurban territories are most likely part of the flight path as well as the habitat of some type of migrating bird. The artist Lynne Hull's ambitious project Migration Mileposts (2000–) not only marks sites of bird migrations with signposts that provide information about the migration routes of various birds but aims to enhance their habitats. Rather than reacting to the birds as avian trespassers, we may embrace a Spinozist "ethics of desire" in which, as Claire Colebrook argues, "affirming one's own becoming is maximized in the affirmation of the becoming of others."[72] Against the model of the home that is static—that is the home as property, as a solid line of defense against the dangers of the other—we can imagine it as a place of becoming, a place in which the becoming of other creatures especially enriches our own possibilities for becoming. Imagine the possibilities for constructing habitats that can support a diverse range of symbiotic relations and interwoven pleasures. Mick Smith notes that environmentalists find wilderness valuable because "it is not entirely dominated, monitored, transformed and constrained

or made to conform to the dictates of its efficient utilization by humans."[73] While terrestrial wilderness areas and marine protected areas
remain crucial for ecosystem and species protection, it is equally crucial for the wild to transgress boundaries, or, ideally, to be invited into
the heretofore human domain.

Public art by Patricia Johanson, Reiko Goto, and Lynne Hull demonstrate possible strategies for integrating human desires with those of
other species. Johanson's Fair Park Lagoon, created in Dallas between
1981 and 1986, transformed a body of polluted water into a habitat
for turtles, fish, and waterfowl by experimenting with bioremediation
methods that were not common in that time (Figure 1). Johanson remembers, "Just before the project was dedicated, flocks of wild birds
arrived. Different species of fish were introduced into an environment
that could nurture them."[74] Xin Wu describes how Johanson based her
design on the shape of a fern: "the elegant shape of the fern would
lend itself to the construction of a cluster of splitting bridges," so
she "transformed each of the slim fronds into bridges according to
architectural building codes."[75] The result is that "people who walk
the paths follow the same curves and rhythms as the biological forms,
repeating the pattern of the plant," registering a "formal correspon-
dence with biological structures . . . first through the senses and the
feet, and only later intellectually."[76] The playful, curving, serpentine
paths dip down into and then out of the water, making for a rather adventurous stroll for humans, especially when the water rises and cov-
ers parts of the path and one must decide whether to retreat, wade, or
leap across. Johanson has said that "the most important aspect of her
art is in the parts" she does "not design."[77] Her work invites the corporeal experience of the unexpected: "Body movement and gardens
of unplanned experience turn spectators into participants, ensuring
both a creative response and some consideration of forces that affect
the landscape and our own lives. I have become increasingly interested
in landscapes that confront us with the world as it exists, rather than
those that think only in anthropocentric and aesthetic terms, which is
ultimately not to our benefit."[78]

Multi-species art aims to please plants and nonhuman animals, and
in the process of doing so may diverge from human aesthetic expec-
tations. Take Goto's 1997 Cho-en (Butterfly Garden) in San Francisco,
which she filled with the "weeds" that the butterflies need to survive.
She encourages those who remark that there is nothing to see in such a

Figure 1. Patricia Johanson, Fair Park Lagoon (1981–1986), Dallas, Texas.
Courtesy of Patricia Johanson.

place to "tour" the undergrowth, "belly down and nose-in-the-dirt."[79] Goto and her partners "envision for the future a wildlife corridor of interconnecting green space designed with habitat needs in mind. It will promote the propagation and movement of wild creatures, and encourage the awareness of Nature's wonders, even within the heart of the city." Goto's 1991 project, No Wall, No Roof, No Anything My House, Doesn't Get Wet, Doesn't Get Blown Down, takes its title from a haiku by the fifteenth-century Zen poet Ikkyu. The timber-framed house, with a pond inside it, "was intended to help support what was left of once prolific tree frog community that had been compromised by aggressive local development."[80] Ironically, the safe haven for the frogs is created in the form of what is destroying them—built human habitats—yet the poetic title opens up human inhabitation by refusing borders—No Wall, No Roof, No Anything—provoking us to imagine multispecies modes of habitation and the often conflicting needs of humans and frogs.

Figure 2. Lynne Hull, Lightning Raptor Roost # 2. Courtesy of Lynne Hull.

Lynne Hull has pioneered "trans-species art": sculptures that appeal to wildlife, such as the Raptor Roost series of sculptures that provide perching and nesting sites for birds of prey (Figure 2). These stark, lightning-bolt-like forms fit perfectly in the vast desert landscape, pleasing the human eye. Hull also hopes the sculptures appeal to the raptor's aesthetic sense and architectural needs: "I hoped the hawks would find them attractive."[81] All these public artworks depend, for their success, on the actions and predilections of nonhuman animals, but it is Hull's comment that reveals how radically art itself is transformed by the notion of animal aesthetics—her Raptor Roosts were created more for the pleasure of hawks than for humans. And who knows? Hawks may be tougher to satisfy than art critics. Hull's biography lists her "clients" as "hawks, eagles, pine martin, osprey, owls, spider monkeys, salmon, butterflies, bees, frogs, toads, newts, bats, beaver, songbirds, otter, rock hyrax, small desert species, waterfowl and occasional humans."[82] As many species face the threat of extinction, Hull envisions artworks as lifeboats: the Polar Platform would provide floating habitat for polar bears, seals, and walruses; the Bird Barge will be a wildlife sanctuary towed in large rivers, which will include plants that "will also help clean harmful substances out of the water."[83] Her website also connects closer to home, as it encourages everyone with a backyard to create trans-species art and sculpture. Clicking on "Get Involved" takes the viewer not to a list of organizations or products to support or purchase by entering credit card numbers, but instead to the material elements that may exist right outside one's own door. She extends an invitation to engage in hospitable projects: "Like People, wildlife need Food, Shelter, Water, and Space to live their lives. By offering of these elements, if you build it, they may come."[84]

An ethics in place can be sparked by the human desire for surprise, for play, for the possibility of becoming, by realizing it is possible for the agency, the activities, the becomings of the nonhuman to re-create a seemingly static site into a place of energy and transformation. Art and architecture that take account of the crossings between human and nonhuman can help us resist the narrow scripting of our lives in which we tread the well-worn paths of work and consumerism. By relinquishing a sense of mastery and instead opening up ourselves and our living, working, and public spaces to the agency, the actions, the memories, and the pleasures of the nonhuman, we can dwell within abundantly inhabited places of transformation. As the theorist Brian Massumi asks, "What is the qualitatively transformative force that makes social ingress? Is it not nature? What is nature 'in itself' if not the world's dynamic reserve of surprise?"[85] Elizabeth Grosz writes that the most "dynamic elements of architecture, as well as those of the arts and social and political life, aspire to revel in the sheer thrill of the unknown."[86] And, as Jeffrey Jerome Cohen notes, with the philosophy of agentism "the human and the nonhuman are granted the ability to forge multiple connections, sustain (or break) transformative relations, to bring about the new thing, to create, to vanish, to surprise."[87] Making space for surprising biophilic pleasures, such as the raptor that approves of the artist's roost, the howl of an urban coyote, the ruckus of raccoons at the bird feeder, or the polite tarantula's visit, may help sustain environmental engagements and fuel modes of inhabiting that invite the play of the world.

2 Eluding Capture

THE SCIENCE, CULTURE, AND PLEASURE OF "QUEER" ANIMALS

◇◇◇

We're Deer. We're Queer. Get Used to It. A new exhibit in Norway outs the animal kingdom.

> —Alisa Opar, "We're Deer. We're Queer.
> Get Used to It"

Biological Exuberance is, above all, an affirmation of life's vitality and infinite possibilities: a worldview that is at once primordial and futuristic, in which gender is kaleidoscopic, sexualities are multiple, and the categories of male and female are fluid and transmutable. A world, in short, exactly like the one we inhabit.

> —Bruce Bagemihl, *Biological Exuberance:
> Animal Homosexuality and Natural Diversity*

We are acting with the best intentions in the world, we want to add reality to scientific objects, but, inevitably, through a sort of tragic bias, we seem always to be subtracting some bit from it. Like a clumsy waiter setting plates on a slanted table, every nice dish slides down and crashes on the ground. Why can we never discover the same stubbornness, the same solid realism by bringing out the obviously webby, "thingy" qualities of matters of concern?

> —Bruno Latour, "Why Has Critique Run Out of Steam?
> From Matters of Fact to Matters of Concern"

Western, Euro-American thought has long waged "nature" and the "natural" against LGBTQ peoples, as well as women, people of color, the colonized, and indigenous peoples. Just as the pernicious histories

of social Darwinism, colonialism, primitivism, and other forms of scientifically infused racism have incited indispensable critiques of the
intermingling of "race" and "nature,"[1] much queer theory has brack-
eted, expelled, or distanced the volatile categories of "nature" and the
"natural," situating queer desire within an entirely social, and very
human, habitat. This sort of segregation of "queer" from "nature"
is hardly appealing to those who seek queer green places. Discussing
the "biopolitical organization of life," Catriona Sandilands argues that
to conceive of "life as queer opens the world to a reading in which
generativity is not reduced to reproductivity, in which the future is
not limited to a repetition of a heteronormative ideal of the Same,
and in which the heterosexual couple and its progeny—or some fac-
simile thereof—are not the privileged bearers of life for ecocriticism."[2]
How the sexuality of nonhuman animals is conceptualized—a curious
subset of "nature," "the natural," or "life," perhaps—may open up
similar readings of the world. The existence of queer animals contests
the Western foundation of heteronormativity as that which came
straight from Nature. The fact that science, cultural theory, and common sense have reacted to the sexual diversity of nonhuman life by
denying, dismissing, closeting, segregating, and otherwise explaining
it away, could entice us to add to rather than subtract from the reality,
as Latour puts it, of queer animals. Queer animals also provoke questions within interdisciplinary theory regarding the relations between
discourse and materiality, culture and nature, mechanistic sex drives
and refined desires, scientific explanation and cultural criticism. As
queer animals are both disclosed by various human knowledge systems and elude capture within those systems, that oscillation serves
up pleasurable and delightful "realities," as well as heaping portions of
epistemological humility, awe, and wonder—essential ingredients for
a less arrogantly anthropocentric anthropocene. Queer animals, as
emergent, agential, and elusive, may provoke an ethical–epistemology
of wonder, as well as a new materialist reckoning with animal pleasure
that releases it from the narrow modernist scripts of genetic determinism, instinctual drives, and, on the flip side, social machinations.
Wonder may be aroused by that which cannot be understood through
simplistic explanations, and pleasure may be inflamed by the sense of
being overcome by the staggering variation and the sheer exuberance
of more-than-human sexualities and genders. Pleasure, impossible to
confine within dichotomies of nature and culture, body and mind,

pulses through an imaginative materiality. As Karen Barad contends, matter "is promiscuous and inventive in its agential wanderings: one might even dare say, imaginative."[3]

Popular science books, such Bruce Bagemihl's monumental *Biological Exuberance: Animal Homosexuality and Natural Diversity* and Joan Roughgarden's *Evolution's Rainbow: Diversity, Gender, and Sexuality in Nature and People*, as well as the work of Myra J. Hird, present possibilities for rethinking nature as "queer," by documenting the many non-human species that engage in or display same-sex sex acts, same-sex child-rearing pairs, intersexuality, multiple "genders," "transvestism," and transsexuality. Bagemihl's 750-page volume, two-thirds of which is "A Wondrous Bestiary" of "Portraits of Homosexual, Bisexual, and Transgendered Wildlife," astounds with its vast compilation of species "in which same sex activities have been scientifically documented."[4] Bagemihl restricts himself to mammals and birds, but even so, he discusses nearly three hundred species and "more than two centuries of scientific research."[5] Rich not only with scientific data, but also with photos, illustrations, and charts, Bagemihl's exhaustively researched volume renders any sense of normative heterosexuality within nature an absurdity. Joan Roughgarden's book, *Evolution's Rainbow: Diversity, Gender, and Sexuality in Nature and People*—which consists of three sections, "Animal Rainbows," "Human Rainbows," and "Cultural Rainbows"—paints an expanse of sexual diversity across both animal and human worlds.[6] The Naturhistorisk museum in Oslo, Norway, opened "the first-ever museum exhibition dedicated to gay animals." "Against Nature?" seeks to "reject the all too well known argument that homosexual behavior is a crime against nature" by displaying species known to engage in homosexual acts. The exhibit "outs" these animals by telling a "fascinating story of the animals' secret life . . . by means of models, photos, texts, and specimens."[7] Ironically, the patriarchal diorama of the early twentieth century that served, as Donna Haraway argues, as a "prophylactic" against "decadence"[8] is followed by an exhibition that unveils sexual diversity in the world of animals. Queer animals have also gained notoriety with the controversy over a German zoo's plan "to test the sexual orientation of six male penguins which have displayed homosexual traits" and set them up with female penguins because they want "the rare Humboldt penguins to breed."[9] After the public outcry, zoo director Heike Kueke reassured people that they would not forcibly break up the homosexual penguin couples, saying, "Everyone can live here as they please."[10] *Dr. Tatiana's Sex Advice to All Creation: The Definitive Guide to the Evolutionary Biology of Sex* includes a letter from a manatee, worried that their son "keeps kissing other males," signed, "Don't Want No Homo in the Florida Keys," responding, "It's not your son who needs straightening out. It's you. Some Homosexual activity is common for animals of all kinds."[11] The television sex show host Dr. Susan Block, with her explicit website, replete with porn videos and sex toys, promotes a peaceful philosophy of "ethical hedonism," based on "the Bonobo Way." (Bonobos, one of two species of chimpanzee in the genus *Pan*, are known for their lavish sexual activity.) Block's "Bonobo Way," which includes a great deal of "lesbian" sex, "supports the repression of violence and the free, exuberant, erotic, raunchy, loving, peaceful, adventurous, consensual expression of pleasure."[12]

According to the website for the "Against Nature?" exhibit, "Homosexuality has been observed in most vertebrate groups, and also from insects, spiders, crustaceans, octopi and parasitic worms. The phenomenon has been reported from more than 1,500 animal species, and is well documented for 500 of them, but the real extent is probably much higher."[13] Notwithstanding the sheer delight of dwelling within a queer bestiary that supplants the dusty, heteronormative Book of Nature, the recognition of the sexual diversity of animals has several significant benefits, starting with a more accurate understanding of nonhuman life. Scientific accounts of queer animals suggest that heteronormativity has damaged and diminished knowledge in biology, anthropology, and other fields. Roughgarden charges that "the scientific silence on homosexuality in animals amounts to a cover-up, deliberate or not," and thus scientists "are professionally responsible for refuting claims that homosexuality is unnatural."[14] Bruce Bagemihl and Myra J. Hird document how the majority of scientists have ignored, closeted, or explained away their observations of same-sex behavior in animals, for fear of risking their reputations, scholarly credibility, academic positions, or straight identities. Most notably, Bagemihl includes a candid reflection of the biologist Valerius Geist, who "still cringe[s] at the memory of seeing old D-ram mount S-ram repeatedly": "I called these actions of the rams *aggrosexual* behavior, for to state that the males had evolved a homosexual society was emotionally beyond me. To conceive of those magnificent beasts as 'queers'—Oh God!"[15] A queer science studies stance parallel to that of feminist empiricism, would insist that the critique and eradication of heteronormative bias will result in a better, more accurate account of the world—simply getting the facts (not so) straight. Although Margaret Cuonzo warns of the possibility for homosexist, anthropocentric, "or even egocentric" bias in accounts of queer animals,[16] these possibilities seem highly unlikely given the pervasive heteronormativity not only in science, but in the wider culture as well.[17] Moreover, as Catriona Sandilands argues, citing the case in which ecologists assumed that the lesbian behavior of seagulls "must be evidence of some major environmental catastrophe" (it wasn't), "the assumption that heterosexuality is the only natural sexual form is clearly not an appropriate benchmark for ecological research."[18] In short, environmental sciences require better accounts of the sexual diversity of natural creatures; otherwise, heteronormative bias may render it even more difficult to understand the effects of various toxicants. Giovanna Di Chiro demonstrates the vital need for environmental sciences and environmental politics that are not propelled by homophobia or misogyny.[19] Endocrine disruptors alone demand an extraordinarily complex and nuanced understanding of the "mangling" (in Andrew Pickering's terms)[20] of environmental science, health, and politics with misogyny, homophobia, and other cultural forces.

From a cultural studies perspective that focuses on discursive contestation, queer animals counter the pernicious and persistent articulation of homosexuality with what is "unnatural." The multitude of examples, given by Bagemihl and Roughgarden, not to mention the explicit photos and illustrations, strongly articulate "queer" with "animal," making sexual diversity part of a larger biodiversity. This cultural studies model of political–discursive contestation, however, may, by definition, bracket all that is not purely discursive—ironically, of course, the animals themselves—and thus limit the possibilities for imagining a queer ethics and politics that is also environmentalist. This difficulty is part of a larger problem within cultural theory of finding ways of allowing matter to matter. But even within the paradigm of discursive contestation, trouble arises, since the normative meanings of "nature" and the "natural" have long coexisted with their inverse: nature as blank, dumb, or even debased materiality. In other words, people bent on damning homosexuals will, no doubt, see all this queer animal sex as shocking depravity, consigning queers to the howling wilderness of bestial perversions. No doubt the rather sweet-looking illustrations of, say, female hedgehog "courtship" and cunnilingus included in Bagemihl's book, which would delight many a gay-affirmative viewer, would disgust others (Figure 3).[21]

Rather than simply toss queer animals into the ring of public opinion to battle the still pervasive sense that homosexuality is "unnatural," we could, instead, clear space for something less rigid and overdetermined than the opposing territories of "nature" and "culture." For cultural critics, who fear that any substantive engagement with nature, science, or materiality is too perilous to pursue, queer animals are segregated into a universe of irrelevance. But it is possible to look to queer animals, not as a moral model or embodiment of some static universal law, but in order to find, in this astounding "biological exuberance," a sense of vast diversity, deviance (in the way that Ladelle McWhorter recasts the term),[22] and a proliferation of astonishing differences that make nonsense of biological reductionism. The sexual activities of nonhuman animals need not be reduced to instinctual drives, but can be understood in more capacious terms, as creative, pleasurable, and sometimes strategic acts within particular animal lifeworlds or "naturecultures."[23]

Epistemology of the Zoological Closet

Eve Sedgwick's paradigm of the "open secret" captures the way in which nonhuman animals have been put in a zoological closet: many have witnessed some sort of same-sex activities between animals and yet still imagine the natural world as unrelentingly straight. Such determined ignorance emerges from a heteronormative epistemology. As Sedgwick explains, ignorance—as well as knowledge—has power: "These ignorances . . . are produced by and correspond to particular knowledges and circulate as part of particular regimes of truth."[24] Decades ago, when my brother was young, my mother bought him a pair of hamsters, choosing two females in order to avoid being overrun by hamster offspring. As it turns out, they engaged, constantly, in oral sex. Despite this memory, I must admit that I was rather astonished by Hird, Roughgarden, and Bagemihl's accounts of the enormous variety of sexual diversity throughout the nonhuman world. Who knew? This sense of astonishment, as I will discuss, below, can rouse a queer green, ethical/epistemological/aesthetic response, even as it may be implicated in regimes of closeted knowledges.

Figure 3. Illustration by John Megahan, which originally appeared in Bruce Bagemihl, *Biological Exuberance.* Courtesy of John Megahan.

The sexual diversity of animals, I would contend, matters. Predominant modes of social theory, however, which still assume a radical separation of nature and culture, tend to minimize the significance of queer animals. Just as much feminist theory has engaged in a "flight from nature,"[25] many cultural critics have cast out queer animals from the field of cultural relevance. Jonathan Marks, for example, in *What It Means to Be 98% Chimpanzee: Apes, People, and Their Genes* takes his place in a long line of people who have attempted to clearly demarcate human from animal by seizing on some key difference: "One of the outstanding hallmarks of human evolution is the extent to which our species has divorced sexuality from reproduction. Most sexuality in other primates is directly associated with reproduction."[26] Just as language, tool use, and other human achievements have been usurped by evidence of similar accomplishments across a range of species, the deluge of evidence of same-sex sex among animals collapses this claim. Marks, however, contends that the female "same-sex genital stimulation" of the bonobo is exceptional, arguing that "virtually all primates are sexually active principally as a reproductive activity."[27] Paul Vasey's extensive studies of Japanese macaques, discussed below, as well as the accounts of hundreds of other species that engage in same-sex pleasures, counter Marks's assertion. More generally, however, Marks criticizes the way we, as humans, look to other primates, especially chimps, as the key to understanding our "true" selves: "They are us, minus something. They are supposed to be our pure biology, unfettered by the trappings of civilization and its discontents. They are humans without humanity. They are nature without culture."[28] On this point, Marks offers a demystifying critique, especially of the way the cultural framework of the scientists may be mistaken as "a contribution of the chimps, rather than for our own input."[29] Notwithstanding Marks's revealing analysis of the epistemological problems that animal ethology poses, the overall effect of his debunking—when unaccompanied by any attempt to formulate productive ways of engaging with scientific accounts of animals—is to banish animals to a wilderness of irrelevance, where they serve as the backdrop for the erection of human sophistication.

Jennifer Terry undertakes a discursive critique of "the scientific fascination with queer animals," in which "animals provide models for scientists seeking to determine a biological substrate of sexual orientation."[30] She exposes how "reproductive sexuality provides the master narrative in studies of animal sexuality and tethers queer animal behavior to the aim of defining reproduction as the ultimate goal of sexual encounters."[31] Drawing on Haraway's work, Terry begins her essay by stating that "animals help us tell stories about ourselves, especially when it comes to matters of sexuality."[32] She concludes by arguing that the "creatures that populate the narrative space called 'nature' are key characters in scientific tales about the past, present, and future. Various tellings of these tales are possible, but they are always shaped by historical, disciplinary, and larger cultural contexts."[33] Terry illuminates such contexts in a useful way throughout the essay. This mode of critique, however, framed as it is by the emphasis on "narrative space," confines animal sexual practices within human stories. Although she serves an important source for Terry, Haraway, especially in her most recent work, seems wary of modes of cultural critique that bracket the materiality and the significance of nonhuman animals. She emphasizes that the concept of the companion species, for example, is not an abstract idea, but emerges from living, historical interactions: "Dogs, in their historical complexity, matter here. Dogs are not an alibi for other themes; dogs are fleshly material–semiotic presences in the body of technoscience. Dogs are not surrogates for theory; they are not here just to think with. They are here to live with."[34] Even as Haraway executed one of the most dazzlingly complex and multidimensional scientific/cultural critiques in her 1989 masterpiece *Primate Visions*, she insisted that the "primates themselves—monkeys, apes, and people—all have some kind of 'authorship.'"[35] Her work on primates and dogs, especially, demonstrates this sort of commitment to them—to the world—even as she admits "how science 'gets at' the world remains far from resolved."[36] It remains challenging to cobble together methodologies that allow for both cultural critique and a commitment to uncovering material realities and agencies.[37] Indeed, such projects must straddle the disciplinary divide between the humanities and the natural sciences.

Cynthia Chris, in *Watching Wildlife*, exposes the heteronormativity of wildlife films, explaining that most "wildlife films posit heterosexual mate selection as not only typical but inevitable and without exception."[38] Even the show *Wild and Weird: Wild Sex* "downplays—even avoids—same-sex behaviors in the cavalcade of animal sexualities it frames as varied."[39] Despite her analysis of the heteronormativity of the wildlife genre, however, Chris ultimately warns against celebrating queer animals: "Evidence of same-sex behaviors among animals and genetic influences on homosexuality among humans is used as ammunition in battles waged over gay rights for which advocates might be better off relying on other discourses through which civil rights are claimed. Such evidence remains inconclusive, uneasily generalizable across species, subject to wildly divergent interpretations, and likely to fail the endeavor of understanding animal behavior on its own terms."[40]

Chris's conflation here of animal sexual behavior with "genetic influences on homosexuality among humans" is disturbing, in that it assumes that if animals do something, they do it because of genetic "programming." The extent to which any sexual orientation could possibly be influenced by genetic factors is a question that is entirely separate from the sexual diversity of animals. Rather than assuming that the "genetic human" is the thing that is equivalent to animality, it would be more accurate to think of animal sex as both cultural and material, and genetics as much more of a dynamic process, inextricably interwoven with organism and environment.[41] While Chris would rather have us "rely on other discourses," in part because the evidence for queer animals is "uneasily generalizable across species and subject to wildly divergent interpretations," I will argue below that this very sense of being "not generalizable" is what makes accounts of animal sexual diversity so potent. They highlight a staggering expanse of sexual diversity in nonhuman creatures that is the very stuff of a vaster biodiversity. Environmentalists and LGBTQ peoples can engage with accounts of the sexual diversity of animals, allowing them to complicate, challenge, enrich, and transform our conceptions of nature, culture, sex, gender, and other fundamental categories.

Roger N. Lancaster in *The Trouble with Nature: Sex in Science and Popular Culture* wades through "a toxic waste dump of ideas," hoping to "discover sophisticated new biological perspectives on sex and sexuality," but encountering instead "the same old reductivism warmed over."[42] He argues that the "attempts at supposedly 'queering' science . . . consolidate an astonishingly *heteronormative* conception of human nature."[43] While he exposes heteronormativity and scientific reductivism, he often does so within the framework of a nature/culture opposition. Such an opposition, of course, underwrites the very reductivism that he condemns.[44] For example, he argues that "society, bonding, hierarchy, slavery, rape, and harem" are "concepts, relations, and activities characteristic of humans" and implies that "facts of nature" and "facts of culture" should remain utterly separate.[45] While "slavery, rape, and harem" may seem too loaded, more neutral terms such as "society, bonding, [and] hierarchy" refer to common characteristics of animal groups. Of course, any human terminology would, to some degree, be a distortion of the practices as they exist within animal culture, and yet to emphasize the problematic transfer of linguistic categories to such an extent that one denies any such characteristics or behaviors to nonhumans would be a mistake.[46] After all, despite the impossibility of perfect translation across human languages and cultures, poetry and slang are still translated. The term "rape, " for example, could be replaced by a less-loaded term such as "forced copulation." But to banish the concept altogether would be to imply that nonhuman animals such as dolphins do not have the capacity to consent and thus only engage in instinctual, not intentional or social, sex acts. Lancaster advocates that we "reject the naturalized regime of heteronormativity in its totality" in order to be "finished with the idea of normal bodies once and for all."[47] Ironically, even though Lancaster's book casts scientific accounts of nature as nothing but "trouble," the surprising range of sexual diversity within nonhuman animals could actually foster his utopian dream of abolishing heteronormativity. Lancaster himself becomes momentarily seduced by Bagemihl's book, which he warns is "anthropomorphic," and "fetishistic," but conjures up "charms and talismans of a coming science that would at least be progressive once again."[48]

When nature and culture are segregated within different disciplinary universes, animal sex is reduced to a mechanistic and reproductive function and human sexuality—in its opulent range of manifestations—becomes, implicitly at least, another achievement that elevates humans above the brute mating behaviors of nonhuman creatures. Rather than closeting queer animals and their cultures within "nature," we can recognize that sex for most species is a mélange of the material and the social, and that queer desire of all sorts is part of an emergent universe of a multitude of naturecultures.[49]

Pursuing Pleasures, Creating Cultures

In contrast to the examples above, which expel queer animals from the social and political, Kim TallBear notes that "indigenous peoples have never forgotten that nonhumans are agential beings engaged in social relations that profoundly shape human lives." Challenging the Western conceptions of nature entails for TallBear an analysis of sexuality, because of their parallel treatment: "Nature and sex have both been defined according to a nature–culture divide. With the rise of scientific authority and management approaches, both sex and nature were rendered as discrete, coherent, troublesome, yet manageable objects."[50] I agree with TallBear's overall assessment here and look forward to her project on "how indigenous stories . . . speak of social relations with nonhumans, and how such relations, although they sometimes approach what we in the West would call 'sex,' do not cohere into 'sexuality' as we know it in Western modernity."[51] It is rather remarkable, however, given the way Western science has generally rendered sex and nature as "manageable objects," that same-sex animal sex seems to provoke a different sort of scientific trajectory in which such activities are not reduced to mechanistic forces or genetically determined instinct, but instead are hyper-culturalized so as to transform them into something that is not at all sexual—or more appropriately, not at all homosexual.

Sex, in nonhumans as well as humans, is partly a learned, social behavior, embedded within, and contributing to, particular material–social environments. Kristin Field and Thomas Waite, for example, begin their study of male guppies with the following premise: "On a longer timescale, social environment and 'learned sexuality' can have dramatic effects on the expression of species-typical sexual behavior."[52] Animals are cultural beings, enmeshed in social organizations, acting, interacting, and communicating. Animal cultures, agencies, and significations animate and overcome the convenient view of "nature" as resource, blank slate for cultural inscription, or brute, mechanistic force. Lest we imagine that the view of animal-as-machine without feelings, sentience, or value vanished with Descartes, Werner Herzog's comments in the documentary *Grizzly Man*, which tag a particular bear as Treadwell's "murderer" at the same time they condemn the "blank stare" of that bear, remind us that the demonization and mechanization of animals persists, even when contradictory.[53] Although sex has been categorized as a biological drive, the recognition of the sheer astonishing diversity of animal "sex-gender" systems[54] provokes us to understand nonhuman animals as "cultural" beings. Bagemihl argues that it is "meaningful to speak of the 'culture' of homosexuality in animals, since the extent and range of variation that is found (between individuals or populations or species) exceeds that provided by genetic programming and begins to enter the realm of individual habits, learned behaviors, and even community-wide 'traditions.'"[55]

The pursuit of pleasure may itself be a dynamic force within some animal cultures. Two of the most prominent markers of "culture," in fact—tool use and language—have arisen, for some animals, as modes of sexual pleasuring. Drawing on the research of Susan Savage-Rumbaugh, which began in the 1970s, Bagemihl describes the "'lexicon' of about a dozen hand and arm gestures[,] each with a specific meaning," that bonobos use to "initiate sexual activity and negotiate various body positions with a partner (of the same or opposite sex)."[56] He includes a chart illustrating these hand movements and translating them into commands such as "Approach" or "Move Your Genitals Around."[57] Bagemihl argues that among primates, humans included, "as sexual interactions become more variable, sexual communication systems become more sophisticated." He concludes that "it is possible, therefore, that sexuality—particularly the fluidity associated with nonreproductive sexual practices—played a significant role in the origin and development of human language."[58]

Bagemihl's claim for the influence of sexuality on the development of tools is equally bold. Citing examples of how many primates not only use, but also manufacture, objects to aid with masturbation, Bagemihl claims that "the pursuit of sexual pleasure may have contributed, in some measure, to our own heritage as creatures whose tool-using practices are among the most polymorphous of any primate."[59] Bagemihl's arguments are compelling, and certainly subvert the grand narratives of the "origins of man," which lay claim to tool-making and language as exclusively human. His claim, however, may still be problematic, in that nonhuman sexual practices become more significant because of their role within linear narratives that culminate in the development of the human. But only a slight shift here is needed to read these examples of tool use and language development as part of particular animal naturecultures in which the pursuit of sexual pleasure is one of the most quintessentially "cultural" sorts of activities. Indeed, it is difficult not to be impressed with the creativity, skill, tenacity, and resourcefulness of a female bonnet macaque who "invented some relatively sophisticated techniques of tool manufacture, regularly employing five specific methods to create or modify natural objects for insertion into her vagina":

> For example, she stripped dry eucalyptus leaves of their foliage with her fingers or teeth and then broke the midrib into a piece less than half an inch long. She also slit dry acacia leaves in half lengthwise (using only a single half) and fashioned short sticks by breaking longer ones into several pieces or detaching portions of a branch. Implements were also vigorously rubbed with her fingers or between her palms prior to being inserted into her vagina, and twigs, leaves, or grass blades were occasionally used unmodified.[60]

An artist at work. It is tempting to read this account through and against Roger N. Lancaster's notion of desire: "This desire is on the side of poetry, in the original and literal sense of the word: *poiesis*, 'production,' as in the making of things and the world. Not an object at all, desire is what makes objects possible."[61] Even though Lancaster places desire "squarely within a social purview,"[62] elaborating an ultra-human sort of sexuality that is all culture and no nature, the toolmaking, language-creating, culturally embedded, pleasurable practices of nonhuman animals penetrate this ostensibly human terrain.

Whereas many cultural critics cast animal sex into the separate sphere of nature, many scientific accounts of queer animal sex have rendered them as entirely "cultural," and thus not sexual. Indeed, Dr. Susan Block's philosophy of the "ethical hedonism" of the bonobo is indicative of a general understanding that the "reason" bonobos have so much sex, including same-sex sex, is to reduce social conflicts. Such explanations make all that mounting seem like just another chore. Whereas Block celebrates the eroticism of the bonobos, many scientific accounts of same-sex genital activities emphasize their social functions in such a way as to define them as anything other than sex. As Vasey explains, much same-sex sexual behavior has been interpreted as "sociosexual," meaning "sexual in terms of their external form, but . . . enacted to mediate some sort of adaptive social goal or breeding strategy."[63] Take, for example, the 1998 textbook *Primate Sexuality* by Alan F. Dixon. The chapter "Sociosexual Behavior and

Homosexuality" begins by making it clear that what might look like same-sex sex among nonhuman primates is merely "motor patterns": "The form and functions of sociosexual patterns vary between species, but the important point is that motor patterns normally associated with sex are sometimes incorporated into the non-sexual sphere of social communications."[64] In order to claim that these "motor patterns" are not sex, he places "sex" in a sphere entirely separate from "social communications," a strange segregation for either hetero- or homosexual relations.[65] Obviously, as Vasey explains, "sexual motivation and social function are not mutually exclusive."[66] "Social function," then, often closets same-sex animal sex, by black-boxing pleasure and elevating the "social" into an abstract and disembodied calculus. The gleeful erotic illustrations appearing in Dixon's textbook, however, counter the reduction of these activities to mechanistic "motor patterns," by depicting several entirely different same-sex primate mounts that, to a less mechanistically constrained eye, suggest such things as desire, effort, playfulness, creativity, pleasure—and sex.

Within this landscape of Byzantine heteronormativity, scientists who do suggest that same-sex genital activity may be something like "sex" often do so tentatively. Meagan K. Shearer and Larry S. Katz state that female goats "may mount other females to obtain sexual stimulation. To the observer, there appears to be a hedonistic component associated with the body pressure and motions involved while mounting."[67] Vasey must put forth a strong case to even begin to claim that the sexual behavior between female Japanese macaques is, in fact, sexual:

> Despite over forty years of intensive research in populations in which females engage in same-sex mounting and courtship . . . there is not a single study in existence demonstrating any sort of sociosexual function for these behaviors. Rather, all the available evidence indicates that female–female mounting and courtship are not sociosexual behaviors. Female Japanese macaques do not use same-sex mounting and courtship to attract male sexual partners, impede reproduction by same-sex competitors . . . , form alliances, foster social relationships outside consortships . . . , communicate about dominance relationships . . . , obtain alloparental

> care . . . , reduce social tension associated with incipient
> aggression . . . , practice for heterosexual activity (i.e.,
> female–male mounting), or reconcile conflicts.[68]

Clearly, same-sex activity between animals is considered "not sex"
until proven otherwise. All possibilities for its existence—other than
pleasure—must be ruled out before it can be understood as sex.[69] The
predominant scientific framework, oddly, parallels the mainstream
environmentalist conception of nature that Sandilands critiques as
"both actively de-eroticized and monolithically heterosexual."[70] As
Sandilands explains, drawing on the work of Greta Gaard, "Eroto-
phobia is clearly linked to the regulation of sexual diversity; normative heterosexuality, especially in its links to science and nature, has

the effect of regulating and instrumentalizing sexuality, linking it to
truth and evolutionary health rather than to pleasure and fulfillment.[71]
Queer animals may play a part, then, in helping us question "ecosexual normativity" through asserting "polymorphous sexualities
and multiple natures."[72] Queer animals may also foster an ontology
in which pleasure and eroticism are neither the result of genetically
determined biological drives nor tools in cultural machinations, but
are creative forces simultaneously emergent within and affecting a
multitude of naturecultures. Pleasure, in this sense, may be understood within Karen Barad's notion of performativity as "materialist,
naturalist, and posthumanist," "that allows matter its due as an active
participant in the world's becoming, its ongoing 'intra-activity.'"[73]

Eluding Capture

The multitude of utterly different models of courtship, sexual activity,
child-rearing arrangements, "gender," "transsexualism," and "transvestism" that Bagemihl and Roughgarden document portray animal
lifeworlds that cannot be understood in reductionist ways. Myra J.
Hird in "Naturally Queer" argues that biology "provides a wealth of
evidence to confound static notions of sexual difference."[74] Her exuberant essay encourages us to imagine *The Joy of Sex* for plants,
fungi, and bacteria": "Schizophyllum, for instance, has more than
28,000 sexes. And sex among these promiscuous mushrooms is literally a 'touch-and-go' event, leading [science writer Jenni] Laidman to
conclude that for fungi there are 'so many genders, so little time.'"[75]

Hird poses queer natures as the quintessential boundary transgressors, rather than assuming that "living and non-living matter" is "the stubborn, inert 'outside' to transgressive potential."[76]

Queer animal sex may de-sediment intransigent cultural categories. For example, Paul L. Vasey and his colleagues, in an investigation of female–female mounting behavior in Japanese macaques, conclude that "female mounting in Japanese macaques is not a defective counterpart to male mounting. There is no evidence that females were attempting to execute male mounts, but failing to do so."[77] Rather, the female mounting was "female-typical," exhibiting a strikingly different repertoire of movements.[78] The macaques may remind us of Judith Butler's contention that homosexuality is not an imitation of heterosexuality, or of J. Halberstam's contention that females have their own versions of "masculinity."[79] Vasey himself argues that his study "raises the much broader issue of what constitutes male or female behavior," since it makes little sense to characterize mounting as "male" when "females, in certain populations, engage in this behavior so frequently, and do so in a female-typical manner."[80] The sex/gender distinction in feminist theory posits gender as a cultural, and thus solely human, construct. Joan Roughgarden, however, sees gender in nonhuman animals, defining it as "the appearance, behavior, and lived history of a sexed body."[81] She notes that "many species have three or more genders," such as the white-throated sparrow, which has "four genders, two male, and two female."[82] These "genders" are distinguished by either a white stripe or a tan stripe, which correspond to aggressive and territorial versus accommodating behaviors. As far as sex goes, 90 percent of the breeding involves a tan-striped bird (of either sex) with a white-striped bird (of either sex).[83] Haraway's call to see animals as other worlds, replete with "significant otherness,"[84] resounds when trying to make sense of the multitude of animal cultures that disrupt human—even feminist, even queer—models.

Just as animal sex (and gender) may complicate the foundations of feminist theory, animal practices may also denaturalize familiar categories and assumptions in queer theory and gay cultures. For one thing, nearly all the animal species, as well as individual animals, that have been documented as engaging in same-sex relations also engage in heterosexual sex, meaning that "universalizing" models of sexuality work better for most nonhuman animals than do "minoritizing" models. The "queer" animals I've been referring to, as a convenient

[handwritten margin note: sexuality relevant trait]

shorthand, are "queer" in a multitude of ways, but rarely do any of them correspond to early twenty-first-century categories of "gay" or "lesbian." Roughgarden explains that most male bighorn sheep live in "homosexual societies," courting and copulating with other males, via anal penetration. It is the nonhomosexual males that are considered "aberrant": "The few males who do not participate in homosexual activity have been labeled 'effeminate' males . . . They differ from 'normal males' by living with the ewes rather than joining all-male groups. These males do not dominate females, are less aggressive overall, and adopt a crouched, female urination posture. These males refuse mounting by other males."[85] As Roughgarden contends, these sheep challenge gay/straight categories: "The 'normal' macho bighorn sheep has full-fledged anal sex with other males. The 'aberrant' ram is the one who is straight—the lack of interest in homosexuality is considered pathological."[86] Inevitably, in an attempt to understand the remarkable differences in animal cultures, most accounts draw on human categories and terms. While she critiques the "biased vocabulary" of scientists, Roughgarden uses many terms lifted too unproblematically from twentieth-century American culture, such as "domestic violence" and "divorce," which flattens and distorts the "significant otherness" of animal cultures.

Interestingly, both Roughgarden and Bagemihl argue that many non-Western cultures have more knowledge of and appreciation for the sexual diversity of the nonhuman world. Roughgarden, for example, notes that in the South Sea islands of Vanuatu, pigs have "been bred for their intersex expressions": "Among the people of Sakao, seven distinct genders are named, ranging from those with the most egg-related external genitalia to those with the most sperm-related external genitalia."[87] Similarly, Bagemihl contends that contemporary theoretical accounts of sexual diversity pale next to both the scientific accounts of animal sexuality and the knowledge systems of particular indigenous groups who recognize animal sexual diversity:

> The animal world—right now, here on earth—is brimming with countless gender variations and shimmering sexual possibilities: entire lizard species that consist only of females who reproduce by virgin birth and also have sex with each other; or some multigendered society of the Ruff, with four distinct categories of male birds,

some of whom court and mate with one another; or
female Spotted Hyenas and Bears who copulate and
give birth through their "penile" clitorides, and male
Greater Rheas who possess "vaginal" phalluses (like the
females of their species) and raise young in two-father
families; or the vibrant transsexualities of coral reef fish,
and the dazzling intersexualities of gyandromorphs and
chimeras. In their quest for "postmodern" patterns of
gender and sexuality, human beings are simply catching
up with the species that have preceded us in evolving
sexual and gender diversity—and aboriginal cultures
have long recognized this.[88]

The rigid heteronormativity of Western culture forecloses such motley, kaleidoscopic bestiaries, whereas more complex sexual and gender
manifestations have been recognized, even esteemed, by some indigenous cultures. Focusing on plants rather than animals, Ana Maria
Bacigalupo's anthropological study *Shamans of the Foye Tree: Gender,
Power, and Healing among the Chilean Mapuche* notes how the Mapuche
valued the exceptional gender fluidity of particular trees and humans,
explaining that during colonial times, "the hermaphroditic *foye* tree
legitimated male *machi*'s co-gendered status as sacred, powerful, and
meaningful." Today the *foye* tree "has become a symbol of office for
both male and female *machi*," or shamans, representing "the *machi*'s
ability to move between worlds, generations, and genders."[89] Rigid
categories, on the other hand, have been the norm for Western scientific reason, as they stroll hand in hand with predilections for domestication, management, and straightforward use. As the above quotation
from Kim TallBear noted, "the rise of scientific authority and management approaches" rendered "both sex and nature" ultimately "manageable objects."[90] While she does not discuss sexual diversity, Celia
Lowe, in *Wild Profusion: Biodiversity Conservation in an Indonesian Archipelago*, warns of the disenchantment of Western knowledge systems:
"Max Weber's famous dictum that instrumental reason disenchants
the world, creating therein an 'iron cage' (what Foucault has called a
'*monstre froid*'), is equally applicable in Indonesia where reason's (imagined) triumph over enchantment has meant that the spirit world itself
has become inhabited by the cold monster of governmental rationality. Compulsory de-magification haunts the postcolonial nation and

the stories it can tell about itself."[91] Disenchantment flattens human encounters with the more-than-human world, limiting knowledge to what is useful. Lowe argues that within the conservation project she studied, "enchantment and disenchantment existed in supplementary relationship; new forms of 'unreason' revealed the limitations of, and aporia in, practices of conservation calculation and management."[92] The fluctuation between enchantment and disenchantment, in Lowe's formulation, yielded more, not less knowledge.

Enchantment and wonder may encourage environmental inclinations. Heather Houser in *Ecosickness in Contemporary U.S. Fiction: Environment and Affect* argues that the task of wonder in the twenty-first century is large: "Wonder must not only shake apathy toward the more-than-human world and move us to curiosity without false idealization; it must also promote concern to curb the destruction of wildlife, of undeveloped space, and of human health and livelihood."[93] Jeffrey Jerome Cohen in *Stone: An Ecology of the Inhuman* writes, "Enchantment is estrangement and secular enmeshment, sudden sighting of the world's dynamism and autonomy, the advent of queered relation."[94] Despite the scientific aim to make sense of the world, to categorize, to map, to find causal relations, many who write about sexual diversity in nonhuman animals are struck with the sense that the remarkable variance regarding sex, "gender," reproduction, and child rearing among animals defies domesticating modes of categorization. These epiphanic moments of wonder ignite an epistemological–ethical sense in which, suddenly, the world is not only more queer than one could have imagined,[95] but more surprisingly itself, meaning that it confounds our categories and systems of understanding.[96] In other words, queer animals elude perfect modes of capture. In Pickering's model, science is "an evolving field of human and material agencies reciprocally engaged in a play of resistance and accommodation in which the former seeks to capture the latter."[97] Paradoxically, this model allows us to value scientific accounts of sexual diversity in nonhuman animals, in the sense that these accounts are accounting for something—something more than a (human) social construction—and yet it also encourages an epistemological–ethical stance that recognizes the inadequacy of human knowledge systems to ever fully account for the natural world.[98]

By eluding perfect modes of capture, queer animals dramatize emergent worlds of desire, action, agency, and interactivity that can never be reduced to a background or resource against which the human defines himself. Donna Haraway, defining "companion species,"[99] explains: "There are no pre-constituted subjects and objects, and no single sources, unitary actors, or final ends. . . . A bestiary of agencies, kinds of relatings, and scores of time trump the imaginings of even the most baroque cosmologists."[100] Such responses emanate from a queer, green place, in which pleasure, desire, and the proliferation of differing lifeworlds and interactions provoke intense, ethical reactions. As Brian Massumi argues, "Intensity is the unassimilable," because, "structure is the place where nothing ever happens, that explanatory heaven in which all eventual permutations are prefigured in a self consistent set of invariant generative rules."[101] Many responses to sexual diversity in nonhuman creatures emanate this sort of intensity of the unassimilable. Volker Sommer, for example, concludes his epilogue to *Homosexual Behavior in Animals: An Evolutionary Perspective* by asking, "Is the diversity of sexual behavior that we can observe in nature anything other than mindbogglingly beautiful?"[102] In a review of Bagemihl's book, Duane Jeffery comments that "nature's inventiveness far outruns our meager ability to categorize her productions," adding that "the sheer inventiveness—exuberance—of nature overwhelms."[103] Joan Roughgarden, herself a transgender woman and ecologist, notes that in writing her book she "found more diversity than [she] had ever dreamed existed," calling her book the "gee-whiz of vertebrate diversity,"[104] an expression that captures the reader's response as much as the book's content. Bagemihl carefully wraps up his "labor of love" with layers of wonderment. We first encounter the poem "Snow" by Louis MacNeice (which includes the line "World is crazier and more of it than we think"), then two lines from e. e. cummings—"hugest whole creation may be less / incalculable than a single kiss"—both of which stand as epigraphs to the entire volume, then an epigraph to the introduction by Einstein: "The most beautiful thing we can experience is the mysterious. It is the source of all true art and science. He to whom this emotion is a stranger, who can no longer pause to wonder and stand rapt in awe, is as good as dead: his eyes are closed."[105] A grand, two-page map of "The World of Animal Homosexuality" on the second and third page of the introduction invites us to see the earth as an entirely different place, one populated with a multitude of queer sexualities. Unlike Latour's clumsy waiter whose "nice dishes" crash to the ground,[106] Bagemihl wishes to deliver "'the facts' about animal behavior" as well as "captur[ing] some of their 'poetry'": "In addition to being interesting from a purely scientific standpoint, these phenomena are also capable of inspiring our deepest feelings of wonder, and our most profound sense of awe."[107] The wonder, awe, and pleasure of contemplating the countless modes of nonhuman sexual diversity, which pulse with desire and erotic ingenuity, may generate environmentalisms that are, of course, already fabulously queer.

PART II
Insurgent Exposure

3 The Naked Word

SPELLING, STRIPPING, LUSTING AS ENVIRONMENTAL PROTEST

◇◇◇

The empress [of feminist theory] is wearing too many clothes. Her fashion selections have become increasingly cumbersome, fussy, restrictive, and ostentatious, and I wish she once again would dare to display more sensate flesh.

—Judith Stacey

They stop their trucks because I'm bare-breasted. The poem keeps their attention. I want them to see in me an image of something beautiful, sacred, and vulnerable—just like the Earth.

—La Tigresa

Although Judith Stacey was speaking metaphorically about the need for feminist theory to cast off its burdensome trappings, the start of the twenty-first century has seen feminists, environmentalists, animal rights activists, and peace activists who literally "dare to display more sensate flesh." Stacey Kalish claims that an estimated "50,000 people have participated in at least 91 naked protests around the world."[1] From New York City to Arkansas, from Cape Town to Sydney, naked bodies lie on the beach or the snow, spelling out bare, unadorned slogans such as "no bush," "no war," "no gm," "truth," or "peace." The targets of these protests vary, from the war in Iraq, climate change, and genetically modified foods, to something more local—such as a Costco store in Mexico City that would destroy a historic site. In an action that may have inspired many of the others, Nigerian women forced Texaco to improve their village by "taking over an oil refinery and threatening to disrobe."[2] In Spain, animal rights activists take the place of bulls on the streets. Two days before the first bull run, "compassionate and fun-loving people from around the world" come to Pamplona, Spain, to run, naked, in the "Human Race." The event seeks a "win–win alternative to having a stampede of terrified animals who end up being tormented and slaughtered in the bull ring."[3] Protesting the SeaWorld float in the Macy's Thanksgiving Day Parade, animal rights activists, wearing "nothing but black and white body paint to resemble orcas[, planned to] squeeze into a bathtub outside the midtown Manhattan store . . . to mimic orcas held in captivity."[4] In Austin, Texas, "Austintatious Babe" confronted supporters of open carry gun laws by bearing her breasts.[5] The international feminist group FEMEN, which started in Ukraine, engages in "sextremism," fighting sexual, reproductive, and religious oppression with protests that feature naked breasts and torsos covered with slogans such as "In Gay We Trust."[6]

Occupying public space in such a way as to place one's body on the line has been a vital mode of signifying opposition, with a history that includes the hunger strikes of Gandhi and the suffragettes, ACT UP's street theater, the civil rights protestors who braved police dogs and water hoses, the Capitol Crawl, the Occupy Wall Street movement, and the Black Lives Matter die-ins. Such protests involve hardship and danger in various degrees, but these protesting bodies have rarely been naked. What, if anything, is the significance of naked protesting bodies? Are activists simply capitalizing on the cultural currency of (female) flesh, or are they—quite literally—embodying a mode of protest that dramatizes the political subject as interconnected with the material world? Can feminist theories of corporeality and performance—as "overdressed" as they may be—help us to decode these practices, experiences, and images? And how are these protests implicated in and responding to racialized politics of embodiment?

Naked protests, exposing the "sensate flesh" that Stacey calls for, mushroom at the same time that material feminisms develop. Material feminisms retain the political incisiveness of discursive critique yet open up new avenues of approach to that which is not, by definition, within the purview of the linguistic or textual, namely, human bodies and nonhuman natures.[7] Feminist corporeal theories, such as the work of Elizabeth Grosz, Elizabeth Wilson, and others, seek to re-engage with the materiality of the body. Gloria Anzaldúa's classic text *Borderlands / La Frontera: The New Mestiza* vividly underscores the materiality of place and flesh, as borders traverse them both; moreover, in Ana Louise Keating's reading of Anzaldúa, language and matter animate

Figure 4. Photograph of Bare Witness protest against genetically modified food. Courtesy of Mike Grenville.

each other.[8] While most feminist corporeal theories remain within the demarcation of the human, the naked protests considered here extend human corporeality into actual places, practicing the art of exposure, dramatizing the allied, coextensive permeabilities of human/animal/environment. Extending Gail Weiss's concept of intercorporeality,[9] I propose the term "trans-corporeality," in which human bodies are not only imbricated with one another but also enmeshed with nonhuman creatures and landscapes. Feminist performance theory, particularly that of Peggy Phelan and Rebecca Schneider, will be brought to bear on the way naked protesting, in underscoring human corporeality and the inhuman interchange with the material world, undertakes the critique, subversion, or evasion of the dominant modes of representation and the gendered scenarios of visibility.

Stripping Off

Hal Foster argues that in a mass media age, "images function in a discourse of 'crisis' to reinject a sense of the real into our lives (which is why images of war are so privileged)," and that "such events often seem to be produced in advance as media spectacles (whose importance is judged in terms of 'effect' or 'impact'). One is confronted with the spectacle of events produced so as to be reproduced as images and sold—of a history first scripted, then translated into pseudohistorical simulations to be consumed."[10] Naked protests function to "reinject a sense of the real" even as they are scripted events performed for public consumption. Holding a banner in front of their naked bodies that says "Climate Lies Uncovered," seven protestors occupy the London lobby of Edelman, a public relations firm spinning coal industry plans.[11] Naked protesting expresses not only a sense of urgency but a particular epistemology, which assumes that layers lie and truth must be uncovered. And yet, not unlike Eve Sedgwick's "open secret," the bodies reveal what we all already know but may avoid or deny. Many protestors self-consciously pose the truth as bare, wrapping their flesh in exposed intentions. One protester for world peace in Glastonbury says, "I was nervous, but I wanted to show my naked commitment to the truth."[12] And a headline on the Bare Witness website reads "Kiwi Protestors Spell Out the 'Naked Truth' about GE."[13] The Bare Witness protest against genetically modified foods (Figure 4) plants the bodies in a field, suggesting that food affects human health by putting bodies in direct contact with the land, a tangible image that contests, in its simplicity, the more complicated, ongoing scientific and popular debates about whether GM foods are dangerous and for whom. Against the apparent recognition that the strategy provokes publicity lies the idealistic apparition of the naked body, suggesting a corporeality in common, of the commons, as exposure signifies the need for environmental protection, justice, or peace. Such slogans as "Disrobe to Disarm," for example, link the lack of garments with a lack of armaments, an emptiness full of utopian intent.

Unlike the passive alpha-bodies of activists who lie down to spell out words, La Tigresa (aka Dona Nieto) stands, bares her breasts, and actively "strips for the trees" to save the old-growth forests of the Pacific Northwest of the United States from logging. Her action joins strange bedfellows: pro-sex feminism and a goddess-worshipping strain of ecofeminism; environmental monkey-wrenching and pornography; "sacred sex" and gritty political occupation.[14] Does her action suggest self-commodification for environmentalist ends or a revolt against all propriety "in defense of Mother Earth"? Clearly, while Earth First!, a radical environmental protest group known for its direct actions, enjoys the publicity from La Tigresa's performances, they are not quite sure how to play them. An Earth First! Listserv announcement recounts:

> A small crowd of concerned citizens, fronted by the legendary bare-breasted poet "La Tigresa," successfully stalled timber operations that threaten the world's two tallest trees at a logging site adjacent to Montgomery Woods State Park in Mendocino County. Reciting her "Earth Goddess" poetry topless, as the loggers respectfully listened, La Tigresa was supported by the local residents, mothers with babies, children and dogs, who were all there to express concerns over the environmental impacts of this logging plan. The timber company was delayed long enough for the California Department of Forestry (CDF) inspectors to come out and find violations of the forest practice rules.[15]

What a wholesome striptease! "Mothers with babies, children and dogs" in attendance, loggers "respectfully" listening to the poetry recited by the topless activist. La Tigresa's performances complicate not only the jobs of the loggers but the rhetoric of Earth First!, a rather masculinist environmental organization that vows "no compromise in defense of Mother Earth." Nieto's performance sexualizes Mother Nature, endowing her not only with breasts but with a big bullhorn to accompany them. Nieto's performances are ripe for ridicule, given the long history of trivializing women by sexualizing them. And yet there is something captivating about her performances. La Tigresa challenges the loggers to "peel back the layers of what you call progress," promising, "I am more beautiful naked—take off my clothes."[16] Stripping appeals to an environmentalist desire for an ethical recognition of nature itself, liberating it from the layers of capitalist, bureaucratic, and legal vestments. Notwithstanding the fact that "nature itself" is a partial construction, the ideal of truth as "naked" represents a particular epistemology, and the body of a stripper performing for men could hardly be draped with any more cultural baggage, this chapter will examine whether the exposed flesh within these performances gestures toward an embodied ethics that is intimate with physical places. Even after acknowledging the status of these acts as performance, image, spectacle, and sign, it is possible to imagine that the exposed flesh may embody an ethical recognition arising from a sense of humans as inescapably woven into landscapes, geographies, and places of environmental harm.[17]

Flesh, Race, and the Politics of (White) Exposure

Despite the fact that the naked protestors did assemble in the flesh at a designated time and place, people who encounter these events will experience them on screens, as digital images. The act of stripping down to an elemental being becomes a set of images within an economy of visibility, circulating on the Internet, or in La Tigresa's case, within a documentary film. Citing Jacques Lacan, Peggy Phelan warns that "visibility is a trap": "It summons surveillance and the law; it provokes voyeurism, fetishism, the colonialist/imperial appetite for possession. Yet it retains a certain political appeal."[18] La Tigresa and the alpha-bodies elude such traps by encouraging other-than-specular relations. The strange concatenation of "virtual" and "real" performed by the alpha-bodies in both geographic and digital sites, for example, creates a spectacle that makes consumption an unsatisfying mode of encounter. Environmental theorists query the seeming disjunction between digital media and place-based politics. Timothy Luke, considering how the new informatics reconstitutes both "nature" and "humanity," asks, "If human beings actually become fully invested in bitspace as their most decisive key environmental niche, then what must environmentalism as a political project become?"[19] Arturo Escobar asks, "Is it possible for women, social movements and others to deploy cyberspatial technologies in ways that do not marginalize place?"[20] Escobar concludes that progressive groups "wishing to appropriate and utilize these technologies for social transformation must build bridges between place and cyberspace."[21] The seeming simultaneity of actual and virtual time/space may enable naked protests to perform the materiality of human bodies and geographical places even as it is apprehended through digital images. The nudity, though difficult to discern, may inject a sense of "the real" into a medium comprising layers and layers of mediation. Richard Grusin contends that the "real is no longer that which is free from mediation, but that which is thoroughly enmeshed with networks of social, technical, aesthetic, political, cultural, or economic mediation."[22]

Enmeshed within networks, the naked figures seem to offer nothing to see. These images may, like nudism, involve "a new way of seeing—almost, in a way, a kind of not-seeing."[23] For when bodies spell out words, the words are recognizable, but the bodies barely so. The bodies in some of the photos look quite white, while other bodies, such as those spelling out "paz" in Argentina,[24] radiate many shades of brown. In many of the photos, however, sexual and racial difference are barely discernible, especially since the viewer is compelled to read the words the bodies spell, not to scrutinize them for shades or shapes of social categorization. Are these bodies posing as "flesh"? Hortense Spillers offers a provocative distinction between "body" and "flesh" in her classic essay "Mama's Baby, Papa's Maybe: An American Grammar Book." Spillers posits that "before the 'body' there is the 'flesh,' that zero degree of social conceptualization that does not escape concealment under the brush of discourse, or the reflexes of iconography."[25] She contends, "The flesh is the concentration of 'ethnicity' that contemporary critical discourses neither acknowledge nor discourse away."[26] It is tempting to draw on this formulation to theorize performances aspiring toward a physicality that circumvents difference in the interest of solidarity. In their prone, exposed, and vulnerable positions, meant, oftentimes, to evoke the sufferings they protest, the naked protestors may be momentarily performing as "flesh," in ways akin to Spillers's conception—"the sheer physical powerlessness" of the body as "thing."[27]

Even though Spillers offers a cogent conception of bodies and flesh, I hesitate to import her formulations into this analysis because her theory emerges from the acute, incomparable suffering of enslaved people. The incommensurability between the two contexts is obvious and excruciating, especially given Spillers's formulation in which "body" and "flesh" correspond to "captive and liberated subject-positions."[28] Naked protestors are obviously neither captive nor abject, as their proximity to the ground is a voluntarily undertaken political performance. Nor do naked protestors, whose message depends in some degree on the trope of the body as wholesome and natural—a trope saturated with straight, white privilege—perform, in Nicole Fleetwood's terms, as "excess flesh." In *Troubling Vision: Performance, Visuality, and Blackness*, Fleetwood explains that "excess flesh" attends to "ways in which black female corporeality is rendered as an excessive overdetermination and as an overdetermined excess." Citing Spillers and Sharon Holland, Fleetwood writes, "Excess flesh is not necessarily a liberatory enactment. *It is a performative that doubles visibility: to see the codes of visuality operating on the (hyper)visible body that is its object.*"[29] Fleetwood's theory, which underscores the codes of visuality, may make an invisible racial horizon—which haunts these scenes—visible, as she contends that "hypervisibility" refers to "both historic and contemporary conceptualizations of blackness as simultaneously invisible and always visible, as underexposed and always exposed."[30] If "excess flesh . . . refracts the gaze back upon itself"[31] then perhaps these scenes of exposed but not "excessive" flesh, made up of predominantly white bodies arranged across landscapes, flatten racial categorization and disrupt codes of visuality by presenting nothing to see. The flattening, in this case, places the human participants on the same level, where no subjects are transcendent and no bodies are excessive. It is true that the viewer is positioned above the scene, but the pleasurable identification derives from an imaginative inhabitation of the bodies in place.

Thus, despite the radical incommensurability and the jarring disjunctions between the historically saturated meanings of flesh, to bracket or avoid the question of race, and to ignore the black feminist writings on historical and performed flesh when interpreting these scenes of exposure, would reiterate the presumption that the human (proper) is white. The naked protestors, insofar as they may be manifesting momentary departures from the humanist human, can be read in the context of black feminist critiques of this hegemonic figure. Alexander Weheliye, in *Habeas Viscus: Racializing Assemblages, Biopolitics, and Black Feminist Theories of the Human*, draws on the writing of Spillers and Sylvia Wynter, asking, "What different modalities of the human come to light if we do not take the liberal humanist figure of Man as the master subject but focus on how humanity has been imagined and lived by those subjects excluded from this domain?"[32] Analyses of racialization from black studies are crucial, Weheliye argues, as they "have the potential to disarticulate the human from Man."[33] Weheliye states that Maurice Merleau-Ponty and Elizabeth Grosz conceptualize flesh as a "vestibular gash in the armor of Man." Flesh, he asserts, gesturing back to Spillers and Wynter, is a "stepping stone toward new genres of the human."[34] Rather than shying away from the question of how these naked protests resonate in terms of race, it may be more productive not to suggest that stripping down excavates or effects a pre- or post-racial world, a convenient white delusion, but to consider, with Spillers, Wynter, and Wehilye, the possibilities for flesh to provoke new "genres of the human." As questions of demographics, immigration, environmental racism, environmental justice, climate justice, and neocolonialisms are internal to environmental politics, the human flesh performed in place may underscore the racial dynamics of social justice in its geopolitical/environmental dimensions. Material feminisms, critical posthumanisms, environmental studies, and other theories under the umbrella of the nonhuman turn may consider whether flesh can provoke new genres of the human and whether different critiques of the humanist human, coming from postcolonial studies, critical race studies, material feminisms, disability studies, animal studies, and other perspectives, can be productively allied. Such questions are beyond the scope of this section, surely. However, even here, multiple, conflicting readings of white exposure become apparent, as the scenes and practices can be interpreted as: an occupation of the flesh that critiques or declines humanist, white transcendence; an appeal to the idealized "natural" body as straight, white, and never abject; or a manifestation of an unmarked universal human that whitewashes its exclusionary histories.[35]

Digital "Thereness" and the Failure of Visibility

The racial dimensions of these protests are difficult to access even if still haunting the scenes, as signification of all sorts is stripped down. As the protestors spell out a simple word or two, the scene becomes strangely silent; the protests seem to have given up on the democratic ideal of debate, dissent, and respectable modes of participation in a political process. A bare demand spelled out on the grass, sand, or snow substitutes for the noisier long-form protest genres—of slogans, chants, signs, or songs. The minimalist discourse extends across a large terrain, quietly but undeniably taking up space. The silence creates space for the tactile, visceral, and proprioceptive, merging body and place but situating the viewer in a rather different position from that of the protestor. The audacious, uncomfortable actions, sand gritty on skin, snow cold and wet, circulate as scenes, as framed landscapes. Although the protesting women in Helena, Montana, in January 2003, for example, "assembled on a cold mountain meadow and lay down on snow and prickly pear cactus to spell the word peace with their naked bodies,"[36] we read the word but cannot discern the prick of the cactus.

Oddly, viewers see the protests from classic perspectives: "Classic perspective orients the field of vision to the viewer's veiled or vanished body, as if the scene itself emanates from the viewer's own gaze. The scene is subservient to that eye, at the same time that eye is erased from implication in the visual field. Within the terms of perspective, there is no reciprocity—the seen does not look back."[37] Unlike the feminist performance artists Schneider describes who perform explicit corporeality while returning or subverting the gaze, these images do not look back at us. The lack of a counter-gaze, a mark of self-objectification, may perform the sense of political disenfranchisement that protestors feel. It may also, however, encourage the viewer to imagine an other-than-specular relation to these protesting bodies, since the viewer's perspective is divorced from the scene. What the viewer wants to see in these photos—a sense of the bodily experience of one's naked body on the sand or snow—cannot be seen. It can only be imagined in a peculiarly visceral, tactile way. Such a moment of corporeal identification, even if only fleetingly felt, may stress how twenty-first-century political subjects are transported through systems of electronic mediation and yet must still consider the substance of particular places in which human and other beings are embedded.

Within the peculiar social space of the Internet, as Diane Saco explains, there is an "experiential interplay and occasional disjuncture between the bodily hereness before the screen and the digital thereness on the screen."[38] In the case of the alpha-bodies, the digital "there" was—is—equally visceral. Thus these images subvert the politics of visibility, pointing us toward an embodied hereness that is, paradoxically, both in the chair of the viewer and simultaneously somewhere else, creating a social space in which a virtual intercorporeality may emerge. Phelan suggests that "perhaps through the ethical acceptance of our failure to be rendered within the terms of the visible, we may find another way to understand the basis of our link to the other within and without our selves."[39] In these photos the failure of visibility catalyzes connections to nonhuman others. As the protestors spell out words in and on places—beaches, fields, or even streets—their flesh coterminous with the ground, they dramatize a corporeality that is not contained by a human frame but extends to nonhuman lives and distant locales.

Tree Stumps on the Periphery

Whereas the distant photographs of the alpha-bodies render the land-scape and the slogan visible but the nude bodies rather hazy, the film about La Tigresa foregrounds her topless torso and places the trees on the periphery. La Tigresa's campy hyper-visibility as "Mother Nature" both parodies and embraces this problematic figure. Departing from much postmodern and poststructuralist feminist theory, which has relentlessly pursued a "flight from nature,"[40] Nieto follows the path advocated by Luce Irigaray: "One must assume the feminine role deliberately. Which means to convert a form of subordination into an affirmation, and thus to begin to thwart it."[41] Nieto also assumes the feminine role of John Berger's pithy formulation: "Men act and women appear. Men look at women. Women watch themselves being looked at. This determines not only most relations between men and women but also the relation of women to themselves. The surveyor of woman in herself is male: the surveyed female. Thus she turns herself into an object—and most particularly an object of vision: a sight."[42] Since most of the film *Striptease to Save the Trees* is shot from behind or beside La Tigresa, the film places the spectator in a particularly female position, though once removed. We do not watch her so much as we watch her being watched. Like other feminist performance artists, however, Nieto undermines the passivity of this scenario by being the author and director of her performance. The film also denies men an unimpeded voyeuristic pleasure. The loggers are well aware that they are being filmed—they cannot just sit back and enjoy the performance. Their watching is being watched. They are self-conscious and uncomfortable. The viewer may wonder whether they are watching a striptease, a pagan ritual, performance art, a poetry reading, or a calculated political action. But the camera's perspective, behind and beside La Tigresa, encourages political alliance rather than a detached or objectifying gaze. The shots of the loggers, on the other hand, are confrontational—head on, penetratingly close. Thus, while Nieto may seem to perform a conventional scenario of female objectification and male viewing, her unmistakable agency and the specular strategies of the film turn these relations inside out.

But where is nature in this specular economy? La Tigresa's performance of her flesh as a metonym for nature complicates Berger's formulation even further. We could revise it to say that "men" act and nature appears. Such a formulation is apt, certainly, within a culture that commits acts of environmental destruction even while venerating paintings, photos, and televised images of Nature (sometimes referred to as "nature porn"). But nature, for the most part, does not watch itself being looked at. Nor does it perform parodies of its own representation. Thus, perhaps it is only through a kind of negativity that its representation can be challenged. As Donna Haraway puts it, "For our unlike partners, well, the action is 'different,' perhaps 'negative' from our linguistic point of view."[43] Interestingly, the trees within *Striptease for the Trees* are not given a starring role. Most of the film is shot on the road, or in the midst of a swath of destruction. The forest is on the edge of the scene. Oddly, then, nature stands as the background to the human drama, hardly noticeable within the film. Yet, rather than read this as an inadvertent example of the kind of pernicious "backgrounding," in Val Plumwood's terms,[44] to which women and nature have long been subjected, it may strategically avoid the perils of visibility of which Phelan warns. For environmentally minded viewers, especially, the film may gesture toward a nature that could somehow escape the surveillance and consumption provoked by visibility. The most prominent shot of a tree, in fact, is a close-up of a large, ragged stump—already fallen prey to its own visible prominence. Perhaps the filmmakers, James Ficklin and K. Rudin, tried not to target the trees by plopping them squarely within the mise-en-scène. Instead, the trees gracefully frame the all-too-human action, existing beside the camera's perspective as an ally. Peggy Phelan has argued that invisibility has the potential for resistance and that performance art "becomes itself through disappearance."[45] The fact that the trees may disappear amplifies the film's political resonance, while their disappearance as image suggests an environmental ethos in which nature exceeds its representations. As Elizabeth Bray and Claire Colebrook argue, "Representation would always remain, in some sense, a negation of matter—a break with a prior materiality."[46] By avoiding the transmogrification of living trees to Mylar image, by declining to give gorgeous views of old-growth forests, the film alerts us to the similarities between aesthetic consumption and the capitalist utilization it protests. Instead, it gestures toward a sense of the living, material trees as nonhuman nature that cannot be contained within human paradigms or representations. In *The Good-Natured Feminist: Ecofeminism and the Quest for Democracy*, Catriona Sandilands explains that "nature cannot be entirely spoken as a positive presence by anyone; any claim to speak of or for nonhuman nature is, to some extent, a misrepresentation."[47] This does not mean that we should give up on representing nature—environmental politics, for one, demands it—but that we foreground the limits of our knowledge: "If the part of nature that is beyond language is to exert an influence on politics, there must be a political recognition of the limits of language to represent nature, which to me means the development of an ethical relation to the Real."[48] On the edge of visibility, on the verge of disappearance, there are possibilities for recognition.[49]

An Ethics of Exposure

Gail Weiss outlines an ethics of intercorporeality in which bodies call us "to respond ethically to one another," noting "our continual inter-actions with other human and nonhuman bodies."[50] When La Tigresa calls to the loggers as the "voice" of the earth—"Every particle of your being has been brought forth from the fiber of my body"[51]—she attempts to provoke an ethical, though visceral, recognition of one's individual human body as comprising the selfsame "stuff" of the bodies of other humans as well as that of nonhuman nature. Intercorporeality cannot be restricted to humans, since an insistence on corporeality as transversal perforates the borders that demarcate the human as such. My term "trans-corporeality"[52] suggests that humans are inter-connected not only with one another but also with the material inter-changes between body, substance, and place. Trans-corporeality casts the human as posthuman, not as a historical progression, but as an assertion that, to echo Bruno Latour, we have never been human—if to be human begins with a separation from, or a disavowal of, the very stuff of the world. Discussing Spinoza, Moira Gatens argues that the body's "identity can never be viewed as a final or finished product as in the case of the Cartesian automaton, since it is a body that is in constant interchange with its environment." She continues, "The human body is radically open to its surroundings and can be composed, re-composed, and decomposed by other bodies."[53] Naked protest drama-tizes this idea, as it stages intimacy between flesh and place.

Metonymic rather than metaphoric relations express a sense of trans-corporeality, a material manifestation of the human body "in constant interchange with its environment." Rather than connecting two unlike, unrelated entities, as metaphors do, metonyms express the slide from like to like, the movement between ultimately inseparable entities. Many of the naked protestors express this sense of their flesh as metonymically related to what they are trying to protect. Lisa Franzetta, the PETA spokesperson, for example, said of a naked protest in Hong Kong, "We're perfectly happy to bare our skin to save the skins of animals exploited for the sake of fashion."[54] This sort of trans-corporeality—in which the skin of the human extends to the skin of the animal—erodes individualist notions of the self as well as transcendent notions of the human. Trans-corporeality, moreover, as an ethical call, emerges from a sense of fleshy permeability. Baring their bodies to the elements, they practice an ethics of exposure, which sets aside the fortification of the "I" in favor of the embrace of the multiple, the intertwined, the sensate. As flesh, substance, matter, we are permeable and, in fact, require the continual input of other forms of matter—air, water, food. The many protests against genetically modified food underscore how the human is embedded within and inseparable from the "environment" that it ingests. Valerie Morse, an organizer for an anti-GM protest, states, "The naked protest was a metaphor for New Zealand, which . . . could be 'stripped bare' by genetic engineering."[55] Although we could read these naked bodies as "symbolizing" the state of the nation, it is more apt to read them metonymically, as literally part of the material/geographic/political place of New Zealand. Performing corporeality as that which is violable entails a political claim against future harm to those bodies, but it also disperses the political subject through risky places where human actions have resulted in landscapes of strange agencies.

And nothing is stranger than climate change, with the effects of too much CO_2 saturating the planet. Take the *Evening Standard*'s headline, on June 12, 2015, "World Naked Bike Ride: Hordes of Nude Cyclists to Descend on London for Protest against Car Culture." The sense of naked vulnerability is doubled in this event; cyclists may be endangered not only by climate change caused in part by car culture, but more immediately, by the cars themselves: the event "is aimed at raising awareness about the safety of cyclists while calling for action on climate change."[56] In 2009, appealing to oenophiles, seven hundred naked activists posed in a French vineyard, each holding a wine bottle aloft, to demonstrate that climate change will affect wine production as well as rainforests.[57] Spencer Tunick photographed this

Figure 5. The U.S. installation artist Spencer Tunick and Greenpeace Switzerland present a living sculpture of hundreds of naked people to symbolize the vulnerability of the glaciers under climate change. Reprinted with permission; copyright Greenpeace / Ex-Pres / Michael Würtenberg.

event as well as an earlier, 2007 Greenpeace protest in which six hundred people stood naked on a glacier in Switzerland to protest climate change (Figure 5). The Greenpeace international website states, "Without clothes, the human body is vulnerable, exposed, its life or death at the whim of the elements. Global warming is stripping away our glaciers and leaving our entire planet vulnerable to extreme weather, floods, sea-level rise, global decreases in carrying capacity and agricultural production, fresh water shortages, disease and mass human dislocations."[58] The volunteers who participated in this "living sculpture" stand naked along the bottom third of the photo, assembled on exposed brown rocks, rendered rather insignificant in comparison to the expanse of melting glaciers that rise behind and above them. The story on the Greenpeace site concludes with this quote from Tunick: "I want my images to go more than skin-deep. I want the viewers to feel the vulnerability of their existence and how it relates closely to the sensitivity of the world's glaciers."[59] The sublime scene, however, may aggravate the disjunction between the scale of embodied human lifeworlds and the scale of geological temporalities. In the video of the event, Tunick says, "What seems hard is not. It's slowly shrinking, slowly melting. The humans will be vulnerable just as the glacier is vulnerable." And later, "It's about the softness of their bodies and the hardness of the glacier and how this glacier is weakening."[60] Vulnerability is paradoxical in the anthropocene, as it is these very bodies, soft or not, participating in larger technological and economic systems, that weaken the glaciers, and yet the enormity of collective human agency is countered by the sense of powerlessness that looms large here, as it does within nearly any other climate change scene.

In the photograph of the Swiss Greenpeace event, the activists look in many different directions; they are not a solid block or unified force. The lack of a confrontational gaze underscores the vulnerability of positioned flesh. The prone, bare bodies, ironically, surrender as a mode of struggle. Repudiating their connection to the warriors that would kill in their name, some of these bodies express their sympathies with the "casualties" of war by playing dead. The website Baring Witness proclaims, "Our exposure of the vulnerable human flesh we all share has created a powerful statement against the naked aggression of our country's policies."[61] Similarly, Wendy Tremayne, a participant in naked protests, argues that the reason why so many women partake in naked protesting is because the naked female body bespeaks more of a sense of "vulnerability" that "maybe connects to people's compassion."[62] Suzanne Hart, another participant, titles her book about the experience *Unreasonable Women Bearing Witness: Naked Vulnerability in the Face of Naked Aggression*.[63] These performances counter and critique the U.S. "style of national masculinity" that, as Bonnie Mann argues, has been enlisted to justify the war in Iraq: "The superpower identity can only be maintained and expressed through repetition, through a staging and restaging of its own omnipotence."[64] Mann contends that even female soldiers—such as the infamous Abu Ghraib prison guard Lynndie England—are given the phallus and "invited to participate in the militarized masculine aesthetic along with the men, to become the one who penetrates the racial other."[65]

Even as queer theory, particularly the discussions of S/M, have complicated understandings of sexuality, power, and gender, a gendered polarity in which the feminine is penetrated by the masculine remains. This sense of being penetrable, as performed by the naked protesters then, may operate as a feminist counter-position to the U.S. masculinist identity of aggression. Anita Roddick's website, for example, features a protest organized by the staff of her daughter Sam's "erotic boutique" in London and includes a rather long philosophical statement about the protest's rationale: "The theme was 'liberate yourself from political bondage' and featured strippers and other sex workers wearing only gas masks and body paints and stencils, delivering a powerful guerilla street performance against war in Iraq . . . The nudity was not simply for shock value. Throughout history, women have used nudity as a poignant symbol of their outrage. The implication of vulnerability is the right symbolism for a world vulnerable to unchecked aggression by the stronger against the weaker."[66]

The photographss feature naked body-painted women who boldly march through the streets of London sporting intimidating gas masks. Fortunately for feminist politics, the proclamation of vulnerability does not preclude a tough, insurgent stance. Many of the accounts of naked protest, in fact, underline the protestors' courage and tenacity. Reporting on a protest in January 2003 in the United Kingdom, the Bare Witness site explains, "They felt compelled to use the shock of their nude bodies to send a message to the government—this is not the way to put an end to war. They could have stayed at home in the warm and shook their heads at the news—but they came out to be vulnerable for peace."[67] Or, more dramatically, "It was a day for wrapping up rather than stripping naked and lying in a field. But thirty residents from Ashdown Forest in East Sussex took off their clothes and lay down to spell out the word 'peace' to show their opposition to attacking Iraq. They obligingly held their poses for what felt like several hours while photographers dilly-dallied with light meters and lenses."[68] John Barry has noted that environmental movements have promoted the sense of "human dependency on nature and related ideas of vulnerability, neediness, frailty, limits, precaution."[69] Whereas Barry argues for "ecological stewardship and virtue," it is important to underscore how both stewardship and virtue reinforce Christian, humanist notions of the individual as a disembodied creature, detached from the environment.

Environmental ethics and politics can instead improvise from within a sense of being embedded, permeable, and profoundly interconnected with climates, landscapes, and nonhuman lives. Astrida Neimanis and Rachel Loewen Walker, for example, in their beautiful essay "Weathering," propose that in order to contest the distant and abstract quality of climate change in Western societies, we "bring climate change *home*" by understanding ourselves as *"weathering"*: "intra-actively made and unmade by the chill of a too-cold winter, the discomfort of a too-hot sun," and "recognizing the multitude of bodies (including our own) that are all co-emerging in the making of these weather-times."[70] Performing exposure can catalyze that very sense of weathering.

Ecoporn and Ecosexuals

The environmental organization Fuck for Forest takes a direct approach to environmental activism, creating a porn site that donates a portion of its funds to saving the rainforest. Their website describes its ecological projects in Mexico, Peru, Brazil, Ecuador, Costa Rica, and Slovakia. The description of one of these projects, "Wild Climax Refuge—Costa Rica" shows a blond man and woman, naked, having sex in the jungle next to a photo of trees in containers, presumably about to be planted as part of the restoration project. Scrolling down, there is nothing erotic, but instead a long description of the rationale for the restoration work and a terrific rant about the rich Americans who go to Costa Rica, buy up huge areas of land, and destroy it.[71]

An article published online at *SF Gate* describes one of the FFF films, playing up the restorative dimension of sex:

> Tommy and Leona are having sex on a tree stump in the middle of a Norwegian clear-cut. Leona, with a mop of dreads and a lip ring, looks dreamily across the demolished forest as Tommy, a little shaggy in nothing but his knit hat, works his magic. A few minutes earlier, Leona and Tommy stood at the same spot lecturing about the evils of industrial forestry. But now they're moaning in feral ecstasy, overcoming the powerful negativity of the place—the broken branches and dried out logs—with the juices of the life force itself.[72]

The "moral" in this film—that sex not only feels good but can also heal the planet—is coherent with the rationales on the website. The Fuck for Forest site overflows with manifestos, rants, and philosophies, connecting sexual freedom and pleasure to environmentalism: "Many times the rulers who are making moral issues against naked bodies and sex are the same fuckers making war and destroying our planet. We wish to get closer to nature by celebrating love and liberty. Fuck for forest or be nude for nature."[73] Fuck for Forest does not hold back. The rather long text of the "Love Manifest," for example, is punctuated with explicit sexual images. While the sex is shown close up, the activism travels far away, suggesting the less than desirable linkages between Western conceptions of sexual liberation and a neocolonial primitivism.

How should we read fucking for the forest or stripping for the trees in terms of feminism? In the video *Striptease to Save the Trees*, La Tigresa, accompanied by a harpist, confronts the loggers, offering them sexual invitations from Mother Nature. Her words simultaneously beckon and deflect, invite and redirect, arouse and bewilder. For example, the rather predictable proclamation "I am the Earth, the cradle of creation" is followed by the enticing (and heretical) "In the creases of my inner thighs lies your salvation."[74] La Tigresa's Mother Nature is reminiscent of that of Emma Goldman, who used the figure of the generously sexual earth as the model for an anarchist culture of pleasure and abundance.[75] Goldman, who often playfully spoke as Mother Earth in her journal of the same name, describes the bounteousness of this anarchist figure: "Mother Earth, with the sources of vast wealth hidden within the folds of her ample bosom, extended her inviting and hospitable arms to all those who came to her from arbitrary and despotic lands—Mother Earth ready to give herself alike to all her children."[76] La Tigresa, similarly, offers herself to the loggers, attempting to redirect their desires. Her performance triangulates desire between herself, the earth, and the loggers, as she attempts to use her own body to persuade them of the pleasures of nature. She calls the loggers to erotically engage with the earth over which they drive: "Abandon your clothes by the riverside. / Stretch your naked body . . . / Press your full legs against my yielding ground / And dip your head into my cool green waters and sip . . ." Despite her seductive invitations, the loggers in the film seem embarrassed and confused rather than aroused. Maybe it's the megaphone she's sporting. Maybe it's the fact that she's chasing them down, shouting her poetry, blocking their bulldozers. Or maybe it is the camera, focusing on their reactions rather than her breasts.

In any case, it is difficult not to be reminded of Luce Irigaray's speculation regarding what would happen if the (female) earth, the ground and background for male subjectivity, bespoke its own agency: "If the earth turned and more especially turned upon herself, the erection of the subject might thereby be disconcerted and risk losing its elevation and penetration. For what would there be to rise up from and exercise his power over? And in?"[77] In La Tigresa's performance, female flesh and silent earth speak—loudly, aggressively, with an unmistakable agenda. As with most contemporary environmental activism, La Tigresa's performance betrays an understandable sense of desperation. But it is playful and parodic as well. As heteronormative as her erotic triangle is, her performance may owe a debt to queer activism, which has profoundly altered protest politics by making pleasure political and the political pleasurable. Gay pride marches, for instance, are seriously fun. The discourse of environmentalism, however, is rarely about pleasure. As Catriona Sandilands explains, drawing on Andrew Ross, it "is not only that abundant pleasure is virtually absent in (most) ecological discourse, but that it is often understood as downright opposed to ecological principles; frugality and simplicity appear to act as antithetical principles to enjoyment or generosity."[78] Rather than preaching "self-limitation and self-denial,"[79] La Tigresa entices with the promise of abundant pleasure. Although her scenes of seduction fall within a heterosexual model, and do not offer alternatives to the "eco-sexual normativity" that Sandilands critiques, La Tigresa performs this role with parodic vengeance. This is hardly the usual scenario of the "lay of the land," since this "land" chases down its supposed lay-er, shouting "seductions" through a bullhorn. As Jeannie Forte explains, "Women performance artists expose their bodies to reclaim them, to assert their own pleasure and sexuality, thus denying the fetishistic pursuit to the point of creating a genuine threat to male hegemonic structures of women. Instead of the male look operating as the controlling factor (as it does with cinema), the woman performance artist exercises control."[80] While La Tigresa is the only speaking voice in the film—with the exception of one amused logger whose comments (unfortunately) are inaudible—it would be overreaching to say she is in control since, alas, the loggers resume their activities. The performance ends; the trees are trucked away. Furthermore, even though La Tigresa seems to take pleasure in her sexual performance, the ostensible political strategy here depends on male heterosexual desire that is aroused and redirected. Not surprisingly, given the long and pernicious history of the figure of Mother Earth, the feminist and the environmentalist aims stand in an uneasy, if not contrary, relation.

La Tigresa's performance may also miss the forest for the trees, in the sense that it is the lack of adequate government regulations and oversight and the unchecked rapacity of the logging industry that is to blame for the destruction of the old-growth forests—not the loggers' lack of desire for Mother Earth. The situation Nieto creates stands, ironically, as a microcosm of wider shortcomings in much environmental discourse, which constructs nature as a place of leisure (for the more wealthy) rather than a place of labor (for the less wealthy).[81] Indeed, La Tigresa barges into the loggers' workplace, converting a place of labor into a place of leisure, as the loggers become spectators. Furthermore, appealing to the loggers as individuals, as autonomous moral agents, simplifies the asymmetrical web of economic relations that encompasses us all. La Tigresa's appeal, however, is not limited to the loggers. A wider audience, who accesses her work through video or the Web, may appreciate her audacity, irony, humor, and, indeed, her pro-sex environmentalism, and may even be provoked to imagine an environmentalism where material needs, human labor, abundant pleasure, and ecological robustness could coexist. Such utopian imaginings are often conjured by performance art. As Phelan puts it,

Peggy Phelan
unmarked

"Performance seeks a kind of psychic and political, which is to say, performance makes a claim about the Real-impossible."[82]

While the online Urban Dictionary defines "ecosexual" as "a person, in a dating sense, who is social and environmentally conscious," and *Gaiam Life* defines it as "those who will date only others of the same environmental persuasion,"[83] the "Ecosexual Manifesto" of the performance artists and activists Annie Sprinkle and Beth Stephens declares that ecosexuals "Make Love with the Earth": "We caress rocks, are pleasured by waterfalls, and admire the earth's curves often." As ecosex activists they vow to "save the mountains, waters and skies by any means necessary, especially through love, joy, and [their] powers of seduction."[84] Their projects and performances include "Sexecological Walking Tours," "Dirty Sexecology: 25 Ways to Make Love to the Earth," and many ecosex weddings, where they declare their amorous commitment to dirt, lakes, coal, snow, and other beloveds in community events around the world. Pleasure, joy, eroticism, and desire are the forces that attract and interpenetrate ecosexual bodies and places. The "Here Comes the Ecosexuals" road trip, for example, addresses disturbing environmental concerns by taking place "where water sources are depleted, damaged, and dammed," including "fracking sites, drought areas, aqueducts, and superfund sites." Nonetheless, Stephens and Sprinkle promise, "This is going to be fun," teasing, "Water makes us wet!"[85] Departing from the somber vulnerability in many naked protests, Stephens and Sprinkle offer playful, erotic, environmental performances in which queer desire is not contained by heteronormativity or the human but affirms sensual, passionate interconnections between humans, trees, rivers, rocks, and more. In so doing, Sprinkle and Stephens extend Sprinkle's earlier feminist performances to an environmental terrain. Linda S. Kaufmann writes, "Annie Sprinkle makes lesbian and bisexual desire, sexuality, and the body substantial—tangible, material, and in your face—sometimes literally and sometimes with a vengeance."[86] One of Sprinkle's "motives for exhibiting her cervix," in her legendary Public Cervix Announcement performance, for example is "because it's fun—and I think fun is really important; because I want to share that with people."[87] On Sprinkle and Stephens's site Sexecology: Where Art Meets Theory Meets Practice Meets Activism, the banner includes a quote from Sprinkle: "We aim to make the environmental movement more sexy, fun, and diverse."[88]

Annie Sprinkle

The naked protests and the performances of La Tigresa, like other forms of radical activism, open themselves to ridicule and dismissal. Tim Ferguson, for example, comments, "It's ridiculous to suggest that we'll change our opinion on Iraq after seeing 1000 nude women forming letters on a hillside, like some soft porn Sesame Street sketch."[89] "Buddha Bear," on the Backpacker website, responds to the news of La Tigresa's bare-breasted performances with the quip, "Oh brother! What will be next, put your axe down, and you'll get a bj?"[90] Despite—or because of—the jokes, however, Nieto has managed to call attention to the fight against old-growth logging. Jane Kay reports that "Jay Leno worked her into a Bill Clinton joke. Rush Limbaugh jeered her on his national radio show."[91] La Tigresa appears in some unlikely places. An article in an online magazine *Anvil* devotes several pages to praising La Tigresa as a master of rhetorical persuasion.[92] Even the Mendocino Redwood Company, oddly enough, includes on its website an article titled "'La Tigresa, Nude Savior of the Forest' Holds FB Rally."[93] Dona Nieto has no small intentions for her activism. She claims, "I've changed some of these guys' lives. But I'd like to change the laws, and I'd like to change history."[94] Similarly, an editorial writer in the *Manila Bulletin* admits that naked protest against the war on Iraq may not stop Bush and Blair, yet he hopes for an "interplanetary trend" in which the war is stopped by "millions of men and women all over the world," including "clerics, evangelists, and nuns (of all faiths, Christian, Muslim, Buddhist, Zen, Taoist, etc.) [who] engage in naked protest."[95] What sort of public sphere would this be where a "striptease for the trees" and global occurrences of naked protests would change history? In an article on the suffragette Mary Leigh, titled "Protesting Like a Girl," Wendy Parkins argues, "Where the specificities of female embodiment have been grounds for exclusion or diminished participation, deliberately drawing attention to their bodies has been an important strategy for women engaged in dissident citizenship. Such dissidents have understood their embodiment not as a limitation but as a means by which the parameters of the political domain could be contested."[96] The naked protests considered here seek to contest "the parameters of the political domain" by exhibiting interdependent, interwoven, human, and nonhuman flesh. Both La Tigresa and the alpha-bodies, significantly, carve out a space for their politics as much as they assert a voice. In doing so, they emphasize actual places, material bodies, and other matters. This emphasis is crucial not only for an environmentalism that must insist on the value of particular places, but—when recast as trans-corporeal space—also for a range of issues, including environmental justice, environmental health, disability rights, and queer politics, that demand a recognition of the coextensiveness of material needs, pleasures, and dangers. Emily Martin has sounded a significant warning about the potential dangers of "fluid bodies"—a cautionary note that is especially relevant to conceptions of trans-corporeality: "To the extent that fluid links among individuals can feed the life of a 'pure' super-individual entity such as the corporation or the planet, our enthusiasm for such links might well be curtailed. We might even begin to feel nostalgic for the blockages of such collective body/persons that were provided by liberal democratic notions of the clearly demarcated individual and modernist notions of the separation between the natural and the human."[97] Notwithstanding Martin's warnings, I would contend that the sense of hyperseparation that is promulgated within the United States is more of a danger, as it distances the individual from any sense of global responsibility and accountability. The discourse of national fear and panic after 9/11 generated a politics of domestic containment as Americans were urged to seal themselves within duct-taped enclosures in the name of freedom (see chapter 1). Fourteen years later, Republican presidential hopeful Donald Trump has won over too many Americans with his blatantly racist anti-immigration discourse and his promise to wall off the border with Mexico. Ensconced in a series of enclosures—from the U.S. borders, to the gated communities or suburbs, to the duct-taped home, to the proper family—the citizen is interpellated as a rigidly bounded individual and, according to Mann, as a masculinist, sovereign entity, the penetrator who is not penetrated (see chapter 4). The Baring Witness movement, on the other hand, encourages a globally aware activism, at least for English speakers who visit the website, which, paradoxically perhaps, insists on embodied persons connected to actual places. The utopian moment in these actions in which "the borders of the controlled, rational, cultivated individual break down"[98] loosens the associations that bind the body to unreason, beastliness, and incivility. This is certainly not to claim that "the body" is utopian and peaceful rather than violent, or that fucking will save the forests, but instead to expand the possibilities for an ethics and politics that do not take the bounded capitalist individual as their starting points. Gail Weiss has argued that one of the "ways that bodies have historically been demoralized . . . is precisely through their exclusion from the 'exalted' domain of morality despite the fact that it is in and through our bodies that we feel the effects of our moral judgments and practices."[99] Performing material bodies as ethical terrains—and, as I would argue, as interconnected with the wider physical landscape—offers possibilities for posthuman ethics.

Judith Stacey's conclusion, that "theory may be more useful as a product of political action than its source,"[100] leads to the question, What sort of feminist theory emerges from these practices of naked protest? For both La Tigresa and the various alpha-bodies, the nakedness exceeds its strategic usefulness as spectacle. While politically effective, certainly, in calling attention to its cause, the naked protests do something more. They embody an urgent sense of conviction, as well as an alternative ethos that acknowledges not only that discourse has material effects but also that the material realm is always already imbricated with the discursive. Disrobing, they momentarily cast off the boundaries of the human, which allows us to imagine corporeality not as a ground of static substance but as a place of possible connections, interconnections, actions, and ethical becomings. Exposing themselves, they dramatize how the material interchanges between human bodies, geographical places, and vast networks of power provoke ethical and political actions. Frustrated with conventional routes for political agency, protestors may seek an ethico-political practice that emerges from a somewhere else that is always already here or there, always palpably embedded within particular material places. Feminist theory, cultural studies, and the environmental humanities can take these seemingly eccentric practices seriously and work toward a more complex and consonant rendition of the flesh we inhabit and the places we are.

4

Climate Systems, Carbon-Heavy Masculinity, and Feminist Exposure

◇◇

The Gendering Climate Change and Sustainability conference poster features the stunning artwork of Kirsten Justesen, a Danish sculptor who uses her body as material. Ice Pedestal Formations #1 (now titled ICE PLINTH #1)[1] displays the naked artist standing on an ice pedestal, bent forward, gloved hands touching the ice by her feet, against a frozen background (Figure 6). The sheer aesthetic power of this image is remarkable—the radiant light, the interplay of blue and white, the translucent yet solid surfaces of ice. The elegance of the image does not detract from its eroticism, as the figure is bent over in a position that is both strong and open to penetration. The Ice Plinth series, which was created in 2000, was preceded by other works involving ice, including the Melting Time series Justesen created in Greenland in 1980, before the recognition of global climate change. She states, "The environmental and political aspect of these works has been growing in proportion to the consciousness of global warming. That was not my intention in 1980."[2] It would be difficult now, however, given the acceleration of climate change, not to read Justesen's Ice Plinth series in that context. Not unlike the naked protesters discussed in the previous chapter, Justesen's performances, which pair melting ice with human flesh, suggest intimacy between person and place. Her nakedness bespeaks human exposure, an openness to the material world in which we are immersed. Justesen has said that her work investigates "meeting points for surfaces using [her] body as a tool."[3] As flesh meets ice it usually recoils, but here, in the stillness of the photos, the human remains in contact with the ice. The contact is more capacious in ICE PLINTH #2, as the figure reaches down to embrace the pedestal, exhibiting protection and care for the ice even while she herself remains within a self-protective posture, the "child's pose," as it is known in

Figure 6. Kirsten Justesen, ICE PLINTH #1, from a series of three, 2000.
Pedestal 60 × 60 × 60 cm. Ice, body, nonskid ice gloves, thermo boots.
Pigment print on canvas 1:1 200 × 200 cm, size variable, edition of 7.
Courtesy of Kirsten Justesen; copyright Kirsten Justesen.

yoga.[4] The thick black boots and gloves punctuate both versions with insurgence and strength.

As the last chapter explored, a sense of precarious, corporeal openness to the material world can be an environmentalist stance. Environmental health and environmental justice movements are propelled by this recognition that one's body is interconnected with material hazards that cannot be externalized. Feminist performance artists and activists have also staged exposure, in performances that are critical, strategic, parodic, or even intentionally revolting (in both senses of the word). Given the parallels and overlaps, it is tempting to collapse environmentalist and feminist performances of exposure, and yet they may resonate differently. Justesen provides an intriguing example. As mentioned above, Justesen's Ice Plinth series is dated with the year 2000, a decade and a half ago, when climate change was not yet a widespread concern. Justesen has been using her body as an artistic medium since the 1960s and has been creating feminist art since the 1970s. She is known for place-based art and body–object relations, but not necessarily for environmental art. Her distinctive and provocative body of work is incredibly rich and varied. It is difficult, now, *not* to read works in the Ice Plinth and Melting Time series, as well as many of her other works that involve ice, in relation to climate change. Even though such readings may deviate from the historical context and artist's intentions, they allow the work to become relevant and provocative in different contexts and frameworks as it is received and encountered through time.

I would like to caution against an interpretive trajectory, however, that forecloses meaning by embedding an environmental stance within feminism. While there are convergences, parallels, and alliances between particular modes of feminism and particular modes of environmentalism, the relation between the two remains an open question, a site for argument, analysis, and contextual specificity, because there are always so many variables—and so many genders and sexualities—as the discussion below will indicate. And yet it remains important, since feminist histories are too often forgotten and feminist scholarship is too often marginalized, to assert the impact and relevance of feminist art, activism, and scholarship. The naked protests in the last chapter—against genetically modified foods, war, animal cruelty, and climate change—include both men and women and are often not explicitly feminist, and yet that does not preclude their feminist and queer lineages as performances of body politics.[5] And here we might falter on the pivot between "feminist" and "feminine," as Justesen's image of the naked flesh on a melting glacier suggests a corporeality that is both tough and precariously exposed. Another case in point: when Spencer Tunick discusses the protest he staged with six hundred people on a glacier, does his use of the term "vulnerability" resonate with a sort of environmentalist femininity, or does it disarticulate vulnerability from the feminine, enabling it to travel across human bodies and environments in an unmarked and thus uninhibited manner? "Without clothes, the human body is vulnerable, exposed, its life or death at the whim of the elements. Global warming is stripping away our glaciers and leaving our entire planet vulnerable to extreme weather, floods, sea-level rise, global decreases in carrying capacity and agricultural production, freshwater shortages, disease and mass human dislocations."[6] There is nothing overtly gendered here, although the history of associating women with vulnerability looms on the edges, even to the extent that one could read the performance as a warning against how climate change may feminize us all (not unlike the threat of estrogenic chemicals).[7] Even though I would like to destabilize the associations between vulnerability and femininity, insurgent performances of exposure nonetheless figure as modes of environmental ethics and politics that may be penetrated with feminist histories. Feminisms can be kept in play by imagining the plinth as a pivot between the vulnerability associated with the feminine and the resolute occupation of a politicized sense of permeability that insists on intersubjective, intercorporeal, and inhuman intimacies.

Justesen's art and the naked protest movements perform an insurgent vulnerability—a recognition of our material interconnection with the wider environment that impels ethical and political responses. An insurgent vulnerability could also be called a politics of exposure,[8] in which the environmentalist recognition of having been exposed to such things as carcinogens, endocrine disruptors, and radiation arouses a political response, which may or may not involve literally exposing oneself, but which does strip off the conventional armor of impermeability that blithe capitalist consumerism requires. Performances of exposure declare that humans are not outside the planet looking in, not floating above the phenomenon of climate change, but instead, that we are always materially interconnected to planetary processes as they emerge in particular places. The sense of substantial interconnection with the world may also motivate a continuing engagement with scientific knowledges that have become even more necessary for the formation of practices and policies that will foster the survival of human and nonhuman life. This chapter will focus on critiquing what the politics of exposure counter and contest: the carbon-heavy masculinities of impenetrability and aggressive consumption and, in another domain, the universalizing modes of detached, scientific vision. But it will also critique some of the feminist responses to climate change, interrogating the sort of feminisms and environmentalisms they articulate. Since a broad terrain of theory is submerged within, rather than rising above, the critiques that follow, it may be worth noting that my analysis is grounded on and indebted to scholarship in feminist epistemology, poststructuralist feminisms, material feminisms, feminist science studies, queer theories, environmental studies, and cultural studies.

The Carbon Footprint of Masculinist Consumerism

A peculiar sort of hypermasculinity of impervious but penetrating subjects has emerged in the United States. Under the Bush regime, the United States was infamous for its swaggeringly dismissive attitude toward climate change. This stance continues to be taken up by various sectors of the American populace. Although Bonnie Mann, in her article "How America Justifies Its War: A Modern/Postmodern Aesthetics of Masculinity and Sovereignty,"[9] does not discuss climate change or other environmental concerns, I suspect the hypermasculine style that she diagnoses has been fueled not only by the pervasive post-9/11 fear of terrorist attacks, but also by a lurking, though repressed, dread of climate change and other environmental disasters. Such a posture, or as Mann puts it, such a "style" of masculine, impenetrable aggression, has been evident in Bush's long refusal to acknowledge the threat of climate change. But the desire for hypermasculine "hard bodies," in Susan Jeffords's term,[10] has also emerged as a consumer phenomenon that has increased U.S. carbon emissions. If, as Jeffords argues, the "indefatigable, muscular, invincible masculine body became the linchpin of the Reagan imaginary,"[11] a similar, rigidly masculine corporeality characterizes the Bush Jr. era, a nationalistic stance of impenetrable masculinity that serves only to exacerbate the climate crisis. The fact that so many U.S. citizens are enthralled by presidential candidate Donald Trump's swaggering, aggressive masculinity—an impenetrable masculinity that is inseparable from racism, xenophobia, and misogyny—demonstrates that the "style" Mann critiques remains potent.

I live in the belly of one of the most ravenous, least sustainable beasts—not just the United States, but Texas. It is well known that the United States gulps down far more than our share of fossil fuels. In Texas, especially, it is difficult to ignore the parodically hypermasculine modes of consumerism in which bigger and harder is better. "McMansions" mushroom as suburban and exurban sprawl devour formerly open spaces and agricultural land. Since 1950 the average number of people living in a U.S. house has diminished by one person yet the size of the average American house "has more than doubled."[12] The architect Ann Surchin points to a fearful populace: "No one knows when the next 9/11 will happen. And these houses represent safety—and the bigger the house, the bigger the fortress."[13] Even more noticeable, perhaps, is the fact that SUVs and pickup trucks have not only grown ludicrously huge but are armed with aggressive impenetrability, covered, as they often are, with armor-like accouterments including big, rugged grille guards and hubcaps arrayed with frightening metal cones that look like medieval weapons. Some of these vehicles sport large metal testicles that hang from the trailer hitch (the hitch itself becomes the penis in this ensemble). One website hawks a special set that was modeled after a "real brahma bull." The photo displays the well-endowed truck next to a Texas longhorn.

Just in case this is all too subtle, the "rolling coal" movement drives in to blatantly connect masculinity, the exuberant production of pollution, and the rejection of environmentalism. People in this movement, usually men, equip their trucks to use more, not less, gasoline, in order to blow out black clouds of soot. One YouTube video, titled "Rolling coal on hot babe," shows men purposely engulfing a young woman in a bikini in black soot, for kicks. Jane Evans, who identifies herself as someone from a "land up and over," responds in the page's comments section, "Two F-wits assault a young women with a cloud of deadly toxins. They should be arrested and charged with doing so. BTW: what sort of country allows vehicles that cause this kind of pollution on the roads? They'd be off the road in a flash in my country."[14] A former "coal roller" describes the appeal: "I know a lot of these guys thrive on how much coal they can roll when they're in town next to hybrid cars. . . . It's just a testosterone thing. It's manhood. It's who can blow the most smoke, whose is blacker. The blacker it is, the more fuel you have in your injectors. It was kind of fun."[15] The link between blackness and masculinity is intriguing, given that this is a white working-class movement. But blackness has long been linked with hypermasculinity in racist U.S. history. The police murders of Michael Brown, Eric Garner, Tamir Rice, and countless other African American men, teens, and children beg the question of whether the stereotypes of black hypermasculinity underwrite some of this racist violence. The coal rollers revel in the fantasy of (black) hypermasculinity while remaining safe from its consequences, as their own bodies are up in the truck cab, not on the line. And the painful irony is that African Americans could hardly be categorized as unassailable, given the brutal assaults by police and the pervasive routine modes of institutionalized violence. Less surprising than the racial undertones is that the coal rollers mock climate change and delight in disobeying the EPA. While coal rollers could be seen as an eccentric fringe movement, they epitomize the jacked-up consumerism of the United States as well as the populist conservative stance against government regulations. They may seem marginal, but they manifest perspectives that exist all too close to the center. Less grotesquely, perhaps, transportation within some European cultures may also be gendered. Merritt Polk, in "Gendering Climate Change through the Transport Sector," argues that "the intersections between cars, masculinity, environmental degradation, speed, social exclusions, power, freedom and technology are all manifest within the gendered norms that are prevalent in the transport sector today."[16]

Analyzing transport and overconsumption in terms of gender enables linkages to what Mann calls a "militarized masculine aesthetic." Mann begins her provocative essay by questioning why, despite the fact that all ostensible justifications Bush gave for the U.S. war in Iraq were exposed as being untrue, there has been "no decisive public outcry, no mass demonstrations [that] rock the capital, no credible popular uprising demand[ing] Bush's resignation for initiating a war without reasons."[17] She argues that an aesthetic of the "remaking of an American manhood"[18] garnered support for the war. Discussing the infamous photos of prisoner abuse from Abu Ghraib prison, in particular the photos of U.S. prison guard Lynndie England, Mann argues, "Here the American woman is given the phallus in true (postmodern) democratic form as the military takes up the practice of racialized gender bending. She is invited to participate in the militarized masculine aesthetic along with the men, to become the one who penetrates the racialized other. . . . In this quintessentially modern, quintessentially postmodern military–technological aesthetic, all Americans are part of the hypermasculinized, but dispersed and systematized technomilitary subject of sublime experience."[19] The fact that women may occupy this colonizing "national masculinity" is hardly a liberatory or laudable form of gender transgression, nor a gay-affirmative or feminist stance. Indeed, this shameful moment in U.S. history demonstrates the post-Marxist cultural studies[20] contention that nearly anything can be articulated, or connected, with anything else—here queer practices of a woman taking up the phallus have been enlisted to promote a masculinist, heteronormative nationalism. While I agree with J. Halberstam that "a major step toward gender parity and one that has been grossly overlooked is the cultivation of female masculinity,"[21] I also think that when gender categories are launched into the national imaginary, we need to be concerned about how they are being deployed. A queer stance in one context may be a neocolonialist position of domination in another, as it remains difficult to disentangle masculinity from militarism. Jasbir Puar notes that heteronormativity "is, as it always has been, indispensable to the promotion of an aggressive militarist, masculinist, race- and class-specific nationalism."[22] After 9/11, however, Puar argues in *Terrorist Assemblages: Homonationalism in Queer Times*, a form of "homonationalism" joined a "national heteronormativity" to "extend the project of U.S. nationalism and imperial expansion endemic to the war on terror."[23] Whether or how the Abu Ghraib photos were queer or even sexual is a complicated question, especially since, when the photos were widely distributed, they put "innocent" U.S. viewers of the news in the position of consuming (enjoying behind a facade of "shock"?) scenes of torture as pornography or pornography as torture. Conventions dictate that in sex, if not in science, the spectator is implicated. And yet the fantasy of perfect, unimpeded, innocent vision reigns within both neocolonialist and global scientific regimes.

The Meta-science of Climate Change and the "View from Nowhere"

Another form of hegemonic masculinity lurks in the representations of climate change science—the invisible, unmarked, ostensibly perspectiveless perspective. Climate change, as a vast, complex, scientific phenomenon, demands a multitude of mathematical calculations, and not just abstract but virtual conceptualizations. This aspect of global climate change may entrench conventional models of scientific objectivity that divide subject from object, knower from known, and assume the view from "nowhere while claiming to be everywhere equally" that Donna J. Haraway has critiqued.[24] Just when feminist epistemologies and popular epidemiologies are emerging in which citizens become their own scientific experts—within the global campaign against toxins, environmental justice movements, green consumerism, AIDS activism, and feminist health movements—official U.S. representations of global climate change present a view erected by experts. Such an epistemological stance of hyperseparation may enable dismissal or denial of global environmental crises.

The U.S. Environmental Protection Agency website, under the George W. Bush administration,[25] avoids the language of vulnerability, risk, danger, threat, crisis, or harm, preferring the bland, innocuous term "effects," as it casually mentions how rising temperatures are "already affecting the environment." Perhaps these effects will be good, perhaps they will be bad: "The extent of climate change effects, and whether these effects prove harmful or beneficial, will vary by region, over time, and with the ability of different societal and environmental systems to adapt to or cope with the change."[26] By contrast, the World Health Organization begins its discussion of "Climate Change and Human Health" as follows: "Climate change is a significant and emerging threat to public health, and changes the way we must look at protecting vulnerable populations."[27] Even within the section on "Ecosystems and Biodiversity," the EPA avoids taking a position on whether climate change may be a bad thing: "These changes can cause adverse or beneficial effects on species. For example, climate change could benefit certain plant or insect species by increasing their ranges. The resulting impacts on ecosystems and humans, however, could be positive or negative depending on whether these species were invasive (e.g., weeds or mosquitoes) or if they were valuable to humans (e.g., food crops or pollinating insects)."[28] This stance of distant neutrality casts uncertainty not as something for which we need to take precautions but as an ontological state in which all responsibility, all accountability, all values, all risks, are magically erased. Uncertainty in this articulation does not point to the necessity of the precautionary principle, but instead serves as a prelude to apathy. This exemplifies the "social construction of ignorance" in Robert Proctor's terms, the sort of ignorance that has long been manufactured to absolve tobacco and chemical companies from blame: "Controversy can be engineered; ignorance and uncertainty can be manufactured, maintained, and disseminated."[29]

While the threat of global climate change and the U.S. responsibility for this threat is conjured away by a bland, free-floating, pervasive uncertainty, a different sort of impassioned voice heralds "the facts"

about a new technology. Under the heading "Fact Sheet: Earth Observation System Will Revolutionize Understanding of How Earth
Works," the EPA casts its faith in a "system of systems" that would
deliver complete knowledge, the Global Earth Observation System
of Systems (GEOSS):

> But while there are thousands of moored and free-
> floating data buoys in the world's oceans, thousands of
> land-based environmental stations, and over 50 environ-
> mental satellites orbiting the globe, all providing millions
> of data sets, most of these technologies do not yet talk
> to each other. Until they do—and all of the individual
> technology is connected as one comprehensive system
> of systems—there will always be blind spots and scien-
> tific uncertainty. Just as a doctor can't diagnose health
> by taking just one measurement, neither can scientists
> really know what's happening on our planet without tak-
> ing earth's pulse everywhere it beats—which is all over
> the globe.
>   The challenge is to connect the scientific dots—to
> build a system of systems that will yield the science on
> which sound policy must be built.[30]

This document poses scientific uncertainty as a momentary obstacle
that new technology can fix, rather than something that is endemic
to the scientific process and to the nature of interconnected agencies
of human and environmental actions and processes. The analogy of
the doctor "taking the earth's pulse" reduces the planet to the size
of one familiar being—a patient—an object of authoritative inquiry.
Like magic, almost, science transforms entangled material systems,
substances, and agencies into one clear bit of diagnostic data—the
pulse. Interestingly, this webpage offers no visual image of the earth or
of this "system of systems," leaving us to imagine the doctor–patient
scenario or to envision some vast sense of unknowable data finally
brought under control by one overarching perspective. By contrast,
the physicist and historian of science Spencer R. Weart argues that
"the tangled nature of climate research reflects Nature itself. The
earth's climate system is so irreducibly complicated that we will never
grasp it completely, in the way that one might grasp a law of physics."[31]

The EPA website lists "Nine Societal Benefits" of this system, including the idea that it will help us "Understand, Assess, Predict, Mitigate, and Adapt to Climate Variability and Change."[32] The ability to understand, assess, and predict global climate change supersedes the goal of reducing carbon emissions. The importance of this system of systems is exaggerated elsewhere on the site. For example, one of the "Substantial Socio-economic Payoffs" is that "more effective air quality monitoring could provide real-time information as well as accurate forecasts that, days in advance, could enable us to mitigate the effects of poor quality through proper transportation and energy use."[33] Well, it could, maybe, if people had alternative systems of transportation or if the government enforced stronger emissions policies for industry. As it stands, many citizens are simply bewildered by the stern, official air quality announcements informing us that it is a "level orange," "level red," or even an unthinkable "level purple" sort of day. More accurate, high-tech system of systems air quality pronouncements will not provide citizens with cleaner air or more options for less environmentally harmful transport or energy. The ability to render reality into information, rather than to effect material change, is the unspoken aspiration. Perhaps it is no oversight, then, that there is no image of the earth on EPA's climate change website. Nina Lykke and Mette Bryld in *Cosmodolphins: Feminist Cultural Studies of Technology, Animals and the Sacred* argue that in NASA's famous "Blue Marble" photo, "Nature is being reinterpreted and transformed from object of material consumption to virtual-reality object of worship, awe, and aesthetic–spiritual consumption."[34] The generic webpages offered at the Environmental Protection Agency site avoid any invocation of the earth as an object of worship or awe. Instead, science itself, which promises to deliver us utterly disembodied, transcendent, and complete knowledge—the system of systems—is venerated and mystified. Perhaps any visual image on these pages would be a sort of idolatry—a "graven image" that would bring these lofty delusions down to earth.

Interestingly, the Group on Earth Observations (GEO), which was formed in 2002 by the G8, does include an image of the earth, but in this rendition the oceans are green, the landmasses are white, and giant icons that look like game pieces orbit the planet.[35] The icons, adjacent to the globe, labeled "Information for the Benefit of Society," dwarf the earth. The materiality, the substance, the regional diversity, the geographic diversity, the atmosphere and weather patterns—all substance of the earth itself is erased in this image, as it becomes a blank slate for information. (Information, it must be noted, that will benefit "society" but not necessarily ecologies, habitats, or nonhuman creatures.) The blankness of the earth's surface makes it seem as if we are waiting for the GEOSS to bring the earth into being—for scientists to perform an act of creation. The cartoonish game pieces, the technological apparatuses, are the focus here—not the earth itself, which has been transmogrified into data divorced from actual places. This is not an image consonant with Bruno Latour's sense of the "circulating reference."[36] The material substance and agencies are not transferred or circulated but simply superseded or written over. And we may well ask whether any scientific system, even the system of systems, can deliver "Information for the Benefit of Society." Which society, whose society? Which members of society will benefit, how will they benefit, and who will be ignored or harmed? This God's-eye perspective, this triumphant, purified neutrality, erases social and political contestations, economic disparities, and the material processes of the entangled, emergent world. It imagines that science floats above earthly processes as well as cultural, economic, and political systems. The rhetoric of this system of systems exemplifies the ideal of the "unity of science" that Sandra Harding contends still lurks within the political unconscious of modern science: "Its claim that there is just one science that can discover the one truth about nature also assumes that there is a distinctive universal human class—some distinctive group of humans—to whom the unique truth about the world could be evident. However, as feminist and postcolonial thinkers have pointed out, this is no longer a plausible assumption for most of the world's peoples."[37]

Female Vulnerability and the Mastery of Nature

Although hypermasculine consumerism may seem a far cry from transcendent scientific perspectives, they both detach themselves from a sense of exposure—the sort of vulnerability that refuses to disavow our immersion within the material world. Moreover, the globalizing visions of some of the discourses of climate change impose a rather troubling binary between universal (masculine) scientific knowledge and the marked vulnerability of impoverished women. "Vulnerability" has, in fact, become a key term in the risk assessments of climate change, where it enables researchers to identify the risk differentials of various groups and regions.[38] Even as it has been important for scholars and international women's organizations to assess the ways women may be more vulnerable to the effects of climate change, this emphasis on female vulnerability may have detrimental consequences, in that: (1) it results in a gendered ontology of feminine corporeal vulnerability as opposed to the scientific (or masculinist) imperviousness; (2) it may provoke a model of agency that poses nature as mere resource; and (3) it reinforces, even essentializes, gender dualisms in a way that undermines gender and sexual diversity. Even as it is crucial to consider the specifically gendered modes of vulnerability that global climate change may exacerbate, a feminist, queer, and trans-affirmative politics must avoid reinstalling rigid gender differences and heteronormativity. Moreover, it seems commonsensical to expect that climate change advocacy be environmentally oriented, in the sense that it should promote the significance of ecosystems and nonhuman creatures—not as mere "resources" for human use, but as valuable in and of themselves. In the age of the sixth great extinction, it is reprehensible that environmental organizations would not consider multispecies perspectives or act to mitigate threats to biodiversity.

Some feminist organizations recognize both the power and the risk of the term "vulnerability," employing it carefully within a context that does not pose women as victims. WEDO, the Women's Environment and Development Organization, which focuses on climate change, corporate accountability, and governance, charges that gender, "a critical aspect of climate change," "remains largely on the outskirts." Thus the organization contends that "women, as the majority of the world's poor, are among the most vulnerable to the impacts of climate change. They are also critical to climate change solutions. WEDO approaches gender and climate change from many angles to ensure that women are present at all levels and dimensions of climate change policy-making and action."[39] Feminist organizations such as WEDO are careful to complement feminine vulnerability with feminist agency, savvy, and survival strategies, calling for more parity in decision making and leadership. For example, the Fifty-second Session of the United Nations Commission on the Status of Women called attention to the fact that "climate change is not a gender-neutral phenomenon."[40] The commission explains, "Given that climate change disproportionately affects the poor, and that women form the majority of the world's poor, women are among the most vulnerable to the effects of climate change."[41] The report notes that "women are particularly vulnerable to natural disasters such as floods, fires, and mudslides," because many girls grow up without learning to swim or climb trees. As Halberstam notes in *Female Masculinity*, "Excessive conventional femininity often associated with female heterosexuality can be bad for your health." Whereas Halberstam notes that the "passivity and inactivity" associated with femininity can be accompanied by "unhealthy body manipulations from anorexia to high-heeled shoes,"[42] the argument for female masculinities becomes even more potent in the midst of climate change disaster scenarios.

In addition, climate change, according to WEDO, affects women's livelihoods as well as their ability to provide food, water, and fuel for their families. While the first five of the numbered paragraphs stress women's particular vulnerabilities, the eighth point emphasizes women's agency:

> Women are not just victims of climate change; they are also powerful agents of change. Women have demonstrated unique knowledge and expertise in leading strategies to combat the effects of climate change, as well as natural disaster management, especially at the grassroots level. . . . Women play a vital leadership role in community revitalization and natural resource management. Overall, however, women tend to be underrepresented in decision-making on sustainable development, including on climate change, and this impedes their ability to contribute their unique and valuable perspectives and expertise on the issue.[43]

The next ten of the numbered points—the bulk of the document—lays out strategies for making climate change policies less gender-blind, more inclusive, and more equal. This is a comprehensive feminist document, which balances the need to address women's particular vulnerability to the effects of climate change with a strong statement regarding women's agency, skills, and right of participation.

This particular document, however, severs its feminist position from any sort of environmentalism. "Nature" is represented here solely in terms of being a resource for domestic use. Although women's distinctive roles in "natural resource management" are mentioned, absent is the lively interspecies web of respectful relations, cosmologies, and kinships valued by many non-Western cultures. Rendering living creatures and ecosystems as inert resources not only parallels but also enables extractive and exploitative systems of colonization. The sense of nature as mere "resource" for use may be utterly inimical to particular cultures, especially those of indigenous peoples, many of whom may be at particular risk from climate change.[44] When the document constructs "woman" as a rather monolithic category—despite the different examples—it does so at the expense of cultural difference and biodiversity. For example, the report states, "If current global warming trends continue, there will be a significant depletion of fish stock and the coral reef destruction will result in loss of key marine ecosystems that are central to supporting marine resources which comprise a major source of women's livelihoods in the region."[45] There is little sense that the marine ecosystems or animals are valuable in and of themselves, but instead, they are mentioned only because they support the livelihoods of the women.

"Woman" seems to stake her claims to political agency along a well-worn path of masculine hegemonic subjectivity, as outlined by Luce Irigaray, in which nature serves as the background against which her agency and subjectivity may emerge. Even as ubiquitous Western association between "woman" and "nature" has been for the most part quite detrimental to women, feminists who would also be environmentalists need to forge modes of agency that are not predicated on transcending "nature." Considering the widely accepted predictions that global climate change may cause the extinction of a million of the world's species by 2050, nature (the dynamic, emergent, interconnected world of plants, animals, habitats, and ecosystems) should be at the foreground—not the background—of climate change policy and politics. Critiquing the predominant "resource management approach" in global environmental politics, Christine Bauhardt advocates a "resource politics approach," which should be understood "as a critique of the exploitation of women's labour as a quasi-natural resource, and as a political strategy that analytically and practically combines feminist economics and queer ecologies."[46] Nina Lykke advocates an intersectional feminist analysis that includes the axis of "human/earth others": "We all (including non-human others) are always caught up in multiple intra-acting axes of power which may mutually reinforce each other, but which may also mutually draw in different directions as far as power and interests are concerned."[47]

Mainstreaming Gender Polarities and Heteronormativity

In their introduction to the volume of essays from the Gender and Climate Change conference held in Copenhagen and published in *Women, Gender and Research (Kvinder, Køn og Forskning)*, copies of which were distributed at COP 15, the international climate change summit of 2009, Hilda Rømer Christensen, Michala Hvidt Breengaard, and Helene Oldrup explain the importance of gender mainstreaming for climate change policy: "Gender mainstreaming is a global and flexible strategy aimed at gender equality. It can be understood as a continuing process of infusing both the institutional culture and the programmatic and analytical efforts of agencies with gendered perspectives."[48] Carolyn Hannan, director of the United Nations Division for the Advancement of Women, describes the context and recommendations of the Fifty-second Session on the Commission of the Status of Women's *Gender Perspectives on Climate Change: Issues Paper*, noting the need "to ensure that gender perspectives are integrated into all national policies and programmes on sustainable development, including those focused on mitigation and adaptation strategies, financial arrangements, technology development and capacity-building in the context of efforts to address climate change."[49]

While gender mainstreaming is a crucial goal for environmental policies, gender diversity and sexual diversity cannot be reined into the category of "woman." Intersectional analysis and policy recommendations remain elusive. Some feminist organizations that castigate the gender-blind policies of governing bodies, for example, ignore sexual orientation. The commission's issues paper charges that "there are important gender perspectives in all aspects of climate change" but fails to mention matters of sexual orientation.[50] The Global Gender and Climate Alliance lays out more categories of concern, acknowledging that "the impacts" of global climate change "will be differentially distributed among different regions, generations, age classes, income groups, occupations, and between women and men. Poor women and men, especially in developing countries, will be disproportionately affected."[51] But surely people who are marginalized, denigrated, ostracized, or even criminalized for their sexual orientation or gender identity may be more at risk during a national disaster; they may even be blamed or punished for "causing" the disaster. In the United States it is all too common to hear Christian extremists blame gay people for all sorts of disasters, as they charge that homosexuality incites the wrath of God. Unfortunately, the very emphasis on gender can erase the existence of GLBTQ peoples by sedimenting heteronormative gender roles as universal. For example, the United Nations document "Mainstreaming Gender into the Climate Change Regime" begins, "The UN is formally committed to gender mainstreaming within all United Nations policies and programmes. In all societies, in all parts of the world, gender equality is not yet realized. Men and women have different roles, responsibilities, and decision-making powers."[52] This framing casts "men" and "women" into clear-cut, universal categories; the objective-sounding statement declaring that they "have different roles, responsibilities, and decision-making powers" freezes gender polarities in a way that erases social struggle and contestation as well as foreclosing any space for those who do not, in fact, fit within these rigid and static categories. As Halberstam writes, "The human potential for incredibly precise classifications has been demonstrated in multiple arenas: why then do we settle for a paucity of classifications when it comes to gender? . . . The point here is that there are many ways to depathologize gender variance and to account for the multiple genders that we already produce and sustain."[53]

As climate change brings abrupt weather events, scrambles established migrations, melts arctic regions, and causes other dramatic and subtle transformations, we cannot seek some sort of ontological refuge within stable, heteronormative gender polarities. As the planet becomes incredibly queer, the strange agencies of natural–cultural processes may motivate conservative responses to cling to what is familiar and ostensibly stable, such as the mythical heteronormative dualisms of "man" and "woman." Seeking gender equality by assuming dualisms marked by difference is a perilous proposition. The Canadian document "Gender Equality and Climate Change," for example, asserts universalized gender differences, untempered by intersectional categories of ethnicity, race, class, culture, religion, sexual, or gender orientation: "Women and men experience different vulnerabilities and cope with natural disasters differently; therefore, an increase in the magnitude and frequency of natural disasters will have different implications for men and women."[54] Feminist organizations, which aim for gender mainstreaming within climate science and policy, may inadvertently be mainstreaming gendered heteronormativity and homophobia, as well as rigid, essentialized notions of what "men" and "women" are. In a world of diverse, multiple, and (to refer back to Bruce Bagemihl's term from chapter 2) exuberant genders and sexualities, the dualism of man and woman cannot serve as a ballast against the rapid, even catastrophic alterations of climate change. There is no safe haven—not in the domestic sphere, not in the family, and not in the invocation of the complementary creatures "man" and "woman."

A feminist response to global climate change must challenge not only the ostensibly universal perspective of big science and the hegemonic masculinity of impenetrable, aggressive consumption but also the tendency to reinforce gendered polarities and heteronormativity. It is my hope that environmental organizations, feminist organizations, queer activists, green consumers, climate and climate justice protestors, and ordinary citizens will continue to create and transform modes of knowledge, forms of political engagement, and daily practices that contend with global climate change from positions within— not above—the ever-emergent world. Perhaps it is possible to foster an insurgent vulnerability or a politics of exposure that does not entrench gender polarities but instead endorses biodiversity, cultural diversity, and sexual diversity, and recognizes that we all inhabit trans-corporeal interchanges, processes, and flows. We can engage in practices of revolt and care, protest and pleasure.

PART III

Strange Agencies in Anthropocene Seas

5

Oceanic Origins, Plastic Activism, and New Materialism at Sea

◇◇

Atomic testing. Dead zones. Oil "spills." Industrial fishing, overfishing, trawling, long lines, shark finning, whaling. Bycatch, bykill, ghost nets. Deep sea mining and drilling. Cruise ship sewage. BP. Fukushima. Radioactive, plastic, and microplastic pollution. Sonic pollution. Climate change. Ocean acidification. Ecosystem collapse. Extinction. The destruction of marine environments is painful to contemplate. Having returned from a week on the Gulf of Mexico, where sea life was sparse, I could hardly bear to read Callum Roberts's *The Unnatural History of the Sea*, which describes the staggering abundance of fish and mammals that once inhabited the oceans. Roberts posits that our "collective amnesia" about the profusion of sea life in the past, and our dismissal of "tales of giant fish or seas bursting with life" as "far-fetched," leads us to set our environmental baselines far too low as "we come to accept the degraded condition of the sea as normal."[1] The oceanographer Sylvia Earle notes that since the "middle of the 20th century, hundreds of millions of tons of ocean wildlife have been removed from the sea, while hundreds of millions of tons of waste have been poured into it."[2] Countless species have been overfished to the point of extinction and numerous marine habitats are being destroyed. Rob Stewart's film *Sharkwater* exposes how the market for shark fin soup has resulted in the slaughter of sharks, taking place globally on such a colossal scale that many species of shark may soon be extinct.[3] The destructive practice of trawling, dating back to the fourteenth century, has been joined by deep sea trawling, which disturbs creatures that may be endangered or as yet undiscovered or both, and decimates deep sea coral reefs, some of which are thousands of years old. Long lines, extending for miles across the ocean, luring in birds, mammals, sea turtles, and fish with hundreds or even thousands of baited hooks, result in wide expanses of death and destruction, as the majority of the animals caught are killed then discarded, in order to harvest one particular type of fish. Whether by long lines, trawling, or huge drift nets, industrial fisheries destroy *most* of the catch as "bycatch"—living creatures cast back as lifeless garbage.[4] Jonathan Safran Foer in *Eating Animals* challenges us to imagine "being served a plate of sushi. But this plate also holds all of the animals that were killed for your serving of sushi. The plate might have to be five feet across."[5] He juxtaposes two scenes that are normally severed, the aestheticized, inert food on the plate and the moment of capture, when animal liveliness was quelled by industrialized fishing. But there is another animal in this scene—the human—a pivotal node in the networks of consumption and pollution that destroy ocean ecologies.

Trans-corporeality at Sea?

In *Bodily Natures: Science, Environment, and the Material Self* I argue for a conception of trans-corporeality that traces the material interchanges across human bodies, animal bodies, and the wider material world.[6] As the material self cannot be disentangled from networks that are simultaneously economic, political, cultural, scientific, and substantial, what was once the ostensibly bounded human subject finds herself in a swirling landscape of uncertainty where practices and actions that were once not even remotely ethical or political matters suddenly become so. Trans-corporeality is a new materialist and posthumanist sense of the human as perpetually interconnected with the flows of substances and the agencies of environments. Activists, as well as everyday practitioners of environmental, environmental health, environmental justice, and climate change movements, work to reveal and reshape the flows of material agencies across regions, environments, animal bodies, and human bodies—even as global capitalism and the medical–industrial complex reassert a more convenient ideology of solidly bounded, individual consumers and benign, discrete products. Although the recognition of trans-corporeality begins with human bodies in their environments, tracing substantial interchanges reveals the permeability of the human, dissolving the outline of the subject. Trans-corporeality is indebted to Judith Butler's conception of the subject as immersed within a matrix of discursive systems,[7] but it transforms that model, insisting that the subject cannot be separated from networks of intra-active material agencies (Karen Barad) and thus cannot ignore the disturbing epistemological quandaries of risk society (Ulrich Beck).[8] As a critical posthumanism, trans-corporeality, by insisting on the material inter- and intra-connections between living creatures and the substances and forces of the world, denies human exceptionalism by considering all species as inter-meshed with particular places and larger, perhaps untraceable currents. Trans-corporeality—in theory, literature, film, activism, and daily life—is a mode of ecomaterialism[9] that discourages fantasies of transcendence and imperviousness that render environmentalism a merely elective and external enterprise.

This chapter examines to what extent trans-corporeality can extend through the seas. The persistent (and convenient) conception of the ocean as so vast and powerful that anything dumped into it will be dispersed into oblivion[10] makes it particularly difficult to capture, map, and publicize the flow of toxins across terrestrial, oceanic, and human habitats. Moreover, many marine habitats, such as those in the benthic and pelagic zones,[11] are not only relatively unknown to scientists but are often depicted as "alien" worlds, precluding them from becoming matters of concern.[12] While respecting the extraordinary singularities of ocean habitats and marine animals spawns critical posthumanist or in-humanist modes of thought,[13] the depiction of the depths as alien casts them beyond the reach of the human when, in fact, all marine zones suffer anthropogenic harms. Two different figurations, with divergent ramifications, articulate terrestrial humans with the seas: evolutionary, aquatic origin stories and trans-corporeal tracings of material agencies and far-flung culpabilities. Analyzing the poetry of Linda Hogan; the science writing of Rachel Carson, Neil Shubin, and others; the scholarship of Stefan Helmreich, Mark McMenamin, and Dianna McMenamin; along with the texts, films, and art of plastic pollution activists, this chapter examines narratives and figurations that connect human bodies to the sea. I will argue that even though the long evolutionary arc that ties humans to their aquatic ancestors may evoke modes of kinship with the seas, formulations that end with the human as a completed product of that process conclude too soon. A more potent marine trans-corporeality would submerge the human within global networks of consumption, waste, and pollution, capturing the strange agencies of the ordinary stuff of our lives.

"My Mother Is a Fish": Aquatic Origins of the (Post)Human

William Faulkner's *As I Lay Dying*, a novel in which an impoverished rural family carts their dead mother's body back to her hometown for burial during the dreadful heat of a southern summer, includes a chapter from the child Vardaman's perspective consisting of only five words, floating on an otherwise blank page: "My mother is a fish."[14] Vardaman caught a big fish the same day his mother died then saw his sister cutting up and frying the fish for dinner. This grotesque transference or conflation of the deaths of the mother and the fish contributes to the black comedy of the novel, which exposes human irrationality, psychological defense mechanisms, and the characters' pathetic, tragic, comic, and confused attempts to comprehend their painful and chaotic world. The characters hardly exemplify the humanist ideal of reason that would elevate *Homo sapiens* above other animals. Thus the text may allude to a Darwinian account of the aquatic origins of the human, even if it is rather unlikely that Vardaman has been schooled in evolutionary theory. Yet it would be a mistake to read the novel as in any way posthumanist, for Vardaman's transference of his mother's death onto the fish is merely a psychological response, an error, a literary joke.

If we take Vardaman's statement, "My mother is a fish," as a literal description of human ancestry, however, we are left with the question of whether origin stories can provoke an environmental ethics, or a substantial sense of connection to "alien" aquatic creatures.[15] Most new materialisms deter origin stories. As descendants of postmodernism and poststructuralism, new materialists maintain a critical stance toward foundations and essentialisms, whereas origin stories tend to demarcate one dense, fossilized source for all that follows. Origins often presume or shore up ontological boundaries, delimiting material agencies and possibilities of becoming, as the present has already been formed by the past. New materialisms, on the other hand, stress encounters, inter-action, intra-action, co-constitution, and the pervasive material agencies that cut across and reconfigure ostensibly separate objects and beings. Gilles Deleuze and Félix Guattari's *A Thousand Plateaus* argues for decentralized rhizomatic developments rather than arboristic origins, and features lines of flight, assemblages, and becomings across species.[16] Donna Haraway in *The Companion Species*

Manifesto shifts the focus from the origin or development of any single species, stressing that dogs and humans co-constituted each other through their significant relations across evolutionary time.[17] In *When Species Meet* she asserts that "all actors become who they are in the dance of relating" and "do not precede their relating."[18] Karen Barad, drawing on Niels Bohr's theoretical physics, relishes the sense of the world as a "dynamic process of intra-activity," in which nothing exists that precedes relations. Barad distinguishes her theory of intra-action from that of inter-action, which still "presumes the prior existence of independent entities or relata."[19] The human becomes posthuman in Barad's theory, as we are entangled with the world's dynamic intra-actions. Reason, science, and ethics cannot emerge from some "exterior position."[20] Emphasizing the dual meaning of "mattering," Barad defines ethics as "intra-acting from within and as part of the world in its becoming."[21] If there are no independent entities, then attempts to determine origins could not be corralled into linear narratives but would radiate in innumerable, matted directions. Notwithstanding the possibility of their impossibility, an intra-active account of origins would insist that the (post)human is that which was and continues to be "part of the world in its becoming."

Charles Darwin, exposing the human as a corporeal amalgamation of creatures both at hand and across vast temporal distances, may have given us our first glimpse of the "posthuman," which would not imply something that follows the human, but instead, that the human has always already been precisely that which is jumbled with creatures that are both other than and yet the source of the species. In a letter Darwin cheerfully proclaimed that "our ancestor was an animal which breathed water, had a swim bladder, a great swimming tail, an imperfect skull, and undoubtedly was a hermaphrodite! Here is a pleasant genealogy for mankind."[22] Darwin, in *The Descent of Man*, softens the blow of evolution by telling many a charming and humorous tale demonstrating how the animals that humans would discount, abuse, or revile actually possess various "human" characteristics of curiosity, reason, language, affection, tool use, and the proclivity for religious experiences. Such tales attempt to find a way around the humanism—or disgust[23]—that would impede readers' recognition of the bodily traces of their evolutionary origins in other creatures: "It is notorious that man is constructed on the same general type or model as other mammals. All the bones in his skeleton can be compared with

corresponding bones in a monkey, bat, or seal. So it is with his mus-
cles, nerves, blood-vessels, and internal viscera. The brain, the most
important of all the organs, follows the same law."[24] The word "notorious" marks this conundrum: the fact that "man" is constructed like
other mammals is somehow both accepted and unacceptable, both
obvious and objectionable.

Perhaps little has changed, in that many remain repulsed by the
idea of their own animality, as horror films such as *The Island of Dr.
Moreau*, with its engrossing spectacles of revolting human–animal hybrids, would suggest. And yet physical relatedness may provoke a rich
ethical sense of kinship between human and other animals. Darwin's
term "the community of descent"[25] resonates with ethical provocations. At the very least, anatomical similarity may deny us mental or
spiritual exceptionalism, especially considering that even "the brain,
the most important of all the organs, follows the same law." Moreover,
tracing human origins further back, before mammals had developed,
ultimately leaves us with the "amphibian-like creature," then the "fishlike animal," and finally the "aquatic animal . . . with the two sexes
united in the same individual."[26]

Against the ideological landscape in which kinship between humans and other primates continues to be met with resistance, disgust,
and horror, it may be worth considering what sort of cultural work
evolutionary origin stories featuring fishy mothers or fathers—or,
more appropriately, an intersex aquatic ancestor—could perform.[27]
While the aquatic ancestor of the human has become commonplace
in the United States—think of those footed "Darwin fish" mounted
on the backs of cars—these images denote an opposition to Christi-
anity's dismissal of science, particularly evolution, but do not suggest
concern for actual fish struggling to exist now. The sea is worlds apart
from this ideological skirmish, despite the piscine emblem.

It is not uncommon, however, for writers, scientists, and ocean
conservationists to promote concern for marine ecologies by asserting
the aquatic origin of the human. In Linda Hogan's poem "Crossings,"
the speaker layers three different eras of "crossed beginnings": when
a form of ocean life ventured onto the land, when the ancestor of the
whale traveled back to the water, and when a human child is born with
the "trace of gill slits." The temporal conjunction creates one imaginative space of "crossed beginnings," where the nascent human and
the fetal whale not only encounter each other on the way to what they

will become but also substantially coincide with each other. When the speaker sees a fetal whale she remembers this evolutionary history:

Not yet whale, it still wore the shadow
of a human face, and fingers
that had grown before the taking
back and turning into fin.

In the next stanza the speaker describes the "longing" provoked by remembering the "terrain of crossed beginnings":

when whales lived on land
and we stepped out of water
to enter our lives in air.[28]

This is a poetically rich moment of crossings and kinship, but rather puzzling, since the "we" who "stepped out of water / to enter our lives in air" is already associated with the human. The poem calls to its readers, as part of that "we," to imagine ourselves as the earliest terrestrial creature. In terms of evolutionary chronologies, this makes little sense, as the "we" that "stepped out of water," a tetrapod, would have been the ancestor of the whale as well as the human, and these two journeys, the tetrapod's transition to land and the whale's transition back to water, would have been separated by about 330 million years.

No matter. By dramatizing an encounter between the not-yet-whale and the ancestor of both the human and whale, Hogan collapses time into a space of transformations, where clear and separate lines of descent are overwhelmed by encounters resonating with ever proliferating kinship. As Carl Zimmer puts it in *At the Water's Edge: Fish with Fingers, Whales with Legs, and How Life Came Ashore but Then Went Back to Sea*, "From water to land, and from land back to water: in the history of life, organisms have crossed such seemingly impenetrable boundaries many times."[29] Knowledge of these transitions, he suggests, may incite "a certain kinship with the rest of creation if you happen to find yourself at the ocean floor surrounded by yellowtails and dolphins."[30]

Rachel Carson, in *The Sea around Us*, exalts the sea as the origin of life: "Beginnings are apt to be shadowy, and so it is with the beginnings of that great mother of life, the sea."[31] She notes that the "sea's first children lived on the organic substances then present in the ocean waters, or like the iron and sulphur bacteria that exist today, lived directly on inorganic food."[32] She narrates how, as millions of years pass, the "stream of life grew more and more complex," from "simple one-celled creatures" to sponges, jellyfish, worms, starfish, and plants.[33] While I have long been a critic of the figuration of Mother Earth, Carson's personification of the maternal sea invites an emotional identification with an otherwise dry account of remote eras and events. It also underscores the abundance of the ancient seas: "During all this time, the continents had no life. There was little to induce living things to come ashore, forsaking their all-providing, all-embracing mother sea."[34] While this figuration, problematically, poses the sea as the Angel in the House of evolution, or as a twentieth-century "empty nester," left behind as some of her children move along to higher ground, the narrative is supplanted by a more trans-corporeal sense of connection between the sea and all living creatures. Carson writes:

> When they went ashore the animals that took up a land life carried with them a part of the sea in their bodies, a heritage which they passed on to their children and which even today links each land animal with its origin in the ancient sea. Fish, amphibian, and reptile, warm-blooded bird and mammal—each of us carries in our veins a salty stream in which the elements sodium, potassium, and calcium are combined in almost the same proportions as in sea water. . . . In the same way, our lime-hardened skeletons are a heritage from the calcium-rich ocean of Cambrian time. Even the proto-plasm that streams within each cell of our bodies has the chemical structure impressed upon all living matter when the first simple creatures were brought forth in the ancient sea.[35]

The sea surges through the bodies of all terrestrial animals, including humans—in our blood, skeletons, and cellular protoplasm. In this passage, Carson crystallizes the vast expanses of evolutionary time and space—nearly impossible to fathom—into a form that is already at hand: a form that is in fact ourselves. Significantly, the heritage, or inheritance, here is not exclusively human, but belongs to "fish, amphibian, and reptile, warm-blooded bird and mammal—each of us." While the reader may assume the terms "us" or "our" refers only to the human, the passage itself suggests a broader community of descent.

Neil Shubin's *Your Inner Fish: A Journey into the 3.5-Billion-Year History of the Human Body* is a less mythic account of how the human body carries within it not the sea, exactly, but traces of our fishy ancestry. Shubin, a paleontologist and evolutionary biologist best known for his discovery of the tiktaalik, a close relative of the tetrapod, received some popular attention, including being named ABC News's Person of the Week, appearing on the *Colbert Report*, and hosting a PBS show, *Your Inner Fish*. Shubin's title, *Your Inner Fish*, which plays off the self-help genre that promises personal growth, is oddly apt, since this popularized account of anatomical evolution and scientific discovery actually devolves into a rather anthropocentric self-help manual. Shubin begins by promising, "Ancient fish bones can be a path to knowledge about who we are and how we got that way."[36] Strangely, Shubin calls us to imagine our bodies teeming with aquatic creatures from the past: "There isn't just a single fish inside our limbs; there is a whole aquarium."[37] He argues that the search for human origins should not stop at African hominids, but instead should extend at least to the tiktaalik— the fossilized remains of an intermediate creature between fish and land animals. He overstates his claim, however, erasing gradations between distant life-forms and much closer relatives: "This fossil is just as much part of our history as the African hominids."[38] Nonetheless, there is something compelling about considering our own bodies as encapsulating not just evolutionary but also planetary history:

> If you know how to look, our body becomes a time capsule that, when opened, tells of critical moments in the history of our planet and of a distant past in ancient oceans, streams, and forests. Changes in the ancient atmosphere are reflected in the molecules that allow our cells to cooperate to make bodies. The environment of ancient streams shaped the basic anatomy of our limbs. . . . This list goes on. This history is our inheritance, one that affects our lives today and will do so in the future.[39]

I relish Shubin's argument that such things as the change in atmosphere or the environment of ancient streams profoundly affect who we are. However, two crucial dimensions are missing here. First, when Shubin encapsulates the planetary past in human bodies, he suggests it is only the planet's past—and not its current, or future, conditions—that will "affect our lives today" and "in the future." The material agencies of the present moment, the changes in the atmosphere, the changes in the climate, the acidification of the ocean, the flooding of the environment with thousands of xenobiotic chemicals—are all rendered inert. In his formulation humans embody planetary history, and yet, as completed and complete entities, they stand outside the here and the now. Second, there is a bizarre insistence on the human as the apex of evolution, in that while we may imagine ourselves filled with an entire aquarium of fascinating creatures, those creatures do not unsettle or transform the human but instead reinforce it from within—giving us a rather carnivorous, chronological heft. This makes the entire planetary history "our inheritance." All other species—living, barely surviving, or long extinct—disappear, as the history of air and water become exclusively about "our lives." While Darwin found "grandeur in this view of life" in which "endless forms most beautiful and most wonderful have been, and are being, evolved,"[40] Shubin removes the human from Darwin's tangled bank.

Shubin titles one section "Digging Fossils—Seeing Ourselves," which entraps us in a rather solipsistic universe: wherever we look, wherever we dig, wherever we explore, we ultimately see the human. Nothing eludes this vast net of anthropocentric solipsism. Although Shubin asks, regarding the jellies, "How can we try to see ourselves in animals that have no nerve cord at all? How about no anus and no mouth?"[41] he answers with a rather sketchy analogy: "We may not look much like sea anemones and jellyfish, but the recipe that builds us is a more intricate version of the one that builds them."[42] Focusing on what they "lack," he diminishes the distinctive features of gelatinous creatures. He is not suspended in wonder or contemplation of the jellyfish, but instead proceeds to argue that our common evolutionary origins only demonstrate that it is humans that are "special," "unique," and "extraordinary."[43] If, as some critics have argued, Shubin's book acts as a refutation of intelligent design, it is a rather depressing state of affairs in which Darwin's complex, philosophical, and literary—not to mention scientific—arguments are reduced by this simplistic account, nearly 150 years later.

Even more disappointing, however, is the utter lack of any ecological or environmental vision within Shubin's work. As we witness the sixth great extinction, which may entail the demise of a million species by 2050, as well as the collapse of entire ecosystems, it is bizarre that *Your Inner Fish* does not address the current state of the planet. The book concludes by promising that recent scientific discoveries on "yeast, flies, worms, and, yes, fish tell us about how our own bodies work, the causes of the many diseases we suffer, and ways we can develop tools to make our lives longer and healthier."[44] Ultimately, the entire planetary history is funneled into an upbeat story about longer and healthier human lives. The epilogue simply repeats this message, albeit a bit more poetically: "I can imagine few things more beautiful or intellectually profound than finding the basis for our humanity, and remedies for many of the ills we suffer, nestled inside some of the most humble creatures that have ever lived on our planet."[45] Despite the coziness suggested by the word "nestled," this vision transforms a multitude of living and extinct creatures—all forms of more-than-human life—into a planetary apothecary, a living or fossilized drugstore for the perpetuation of the human. The potential for ethical relations within Darwin's term "community of descent" is short-circuited here, when all life becomes a mere tool for the betterment of *Homo sapiens*. But there is another, more epistemological lack in this book, which relies on a much-critiqued notion of scientific objectivity in which the scientist is the knowing subject and the rest of the world is reduced to inert objects of knowledge. The body that the book conjures, filled with an aquarium of fishes, is not one that anyone actually inhabits; it can be known only through scientific origin stories, not mediated experience.[46] The human reason of the science, which promises to "explain and make our universe knowable," does not itself emanate from the inner fish, or the body as aquarium, or indeed from the body that is, in the current moment, part of the flux and interchanges of the material world. Whereas Darwin, one could argue, forges a scientific and philosophical "posthumanism," in which there are no solid demarcations between human and animal, and in which the human is coextensive with the emergent natural/cultural world, Shubin ultimately offers a much more humanist vision of exceptionalism and containment.

Saturating Terrestrial Life: Aquatic Ancestors, Hypersea, and Genetic Soup

If one of the obstacles for ocean conservation movements is that terrestrial humans are disconnected from the aquatic habitats that cover much of the planet, then narratives, theories, paradigms, and practices that reveal interconnections between these spheres may encourage marine environmentalisms. In Nick Hayes's graphic novel *The Rime of the Modern Mariner*, the mariner encounters jellyfish and squid with bits of plastic "embedded in their flesh." This dreadful spectacle is amplified by the mariner's epiphany: "The thing was like my brother's son . . . / A kin through evolution / A progenitor of mankind / Poisoned by pollution."[47] Such evolutionary narratives are complemented by tropes that enmesh living and nonliving, disclosing how the very water of the oceans flows through human bodies. Sylvia Earle, for example, connects the distant evolutionary past with the immediacy of living human bodies: "Our origins are there, reflected in the briny solution coursing through our veins and in the underlying chemistry that links us to all other life."[48] Julia Whitty laments that although "we carry the ocean within ourselves, in our blood and in our eyes, so that we see through seawater," we nonetheless "appear blind to its fate."[49] In "Human Nature at Sea," the anthropologist and science studies scholar Stefan Helmreich, quoting Earle, the science writer Carl Safina, and the singer Bjork, observes that such "pronouncements cast seawater as a shared substance that makes it possible to feel an embodied human kinship with the aqueous Earth. Environmentally concerned scientists hope that such kinship will lead humans to imagine themselves as linked to the planet both personally and evolutionarily."[50] I agree with Helmreich that such formulations evoke a sense of kinship that marine environmentalists hope will translate into ethical and political commitments to sea life. But as the critique of Shubin's *Your Inner Fish* demonstrates, even though such formulations evoke a "community of descent" across vast temporal and oceanic expanses, their reach may be acquisitive, shoring up human heft, rather than opening out onto the contemporary transcorporealities and tracing human immersion within global networks of harm. Whereas Shubin's excursions ultimately return the solid ground of distancing and disengagement, as the "inner fish" paradoxically epitomizes human exceptionalism, for Carson, Whitty, and

Earle the idea that humans originated from the seas and that we still carry the seas within us situates potential ethical and political recognitions as arising from a trans-corporeal tracing that traverses time and space. By dramatizing the palpable presence of our oceanic origins, they summon these evolutionary traces to catalyze contemporary commitments to the creatures and ecologies that extend across metonymic expanses. In such figurations, ethics begins not with an encounter between self and the other, but with discerning the genealogies, substances, and agencies that diminish the distance between human and sea, as the human becomes more liquid, less solid.

Mark McMenamin and Dianna McMenamin, in their book *Hypersea: Life on Land*, dismiss the idea of blood as seawater by critiquing the film *Hemo the Magnificent*, which "portrayed the saltiness of blood as a legacy of the marine environment of our fishy ancestors."[51] They state that this "lovely" idea "may have helped a lot of people bond with the planet and enjoy the idea of evolution . . . it was a fairy tale," since the blood of vertebrates is, and has been, less salty than the sea.[52] Nonetheless, their book "resurrect[s,] albeit in a modified form, the hypothesis that the sap of plants and the blood of land animals has an evolutionary connection with seawater."[53] They argue not that terrestrial life has its origins in the seas but that it *is* sea life. Contending that their conception of "Hypersea" is both a "physical entity" and a "new scientific theory" they explain, "Land organisms have, by necessity, evolved together as part of a greater interconnected mass of living cells. In moving out of marine waters, complex life has taken the sea beyond the sea and folded it back inside of itself to form Hypersea."[54] Moreover, they stress that "living fluids are not a mere remnant or analog of the sea; they are actually are a new type of sea or marine environment: Hypersea."[55] The idea that the fluids pulsing through living creatures are a kind of marine environment and that all "*plant, animal, protoctistan, and fungal life on land,*" plus their associated "*viral or bacterial symbionts or parasites,*"[56] constitute the "Hypersea," is intriguing. The sea seems to be everywhere, within us, around us, regardless of how arid our terrestrial habitat may be. Yet, if the concept of "Hypersea" blankets nearly everything in the same aqueous composition, the distinctiveness of marine habitats, ecosystems, and creatures is lost even as many new species of ocean life are only just now being found.[57] And while it may seem profoundly posthumanist to envision humans not as distinct individuals but as "reservoirs of Hypersea" that are inhabited by other organisms,[58] the book concludes by claiming that the concept of Hypersea will benefit human health because "many contemporary health threats have a hypermarine aspect—that is, they owe in part to the fact that body fluid is, to a certain extent, a shared resource."[59] Carl Zimmer asks, "What difference will it make to us if Hypersea is real? For one thing, we'll have to realize that we humans are stirring it up like never before. Just as we have brought zebra mussels from Europe to the waters of the United States, we have probably brought diseases like AIDS very quickly from one reservoir of Hypersea (monkeys) to another (ourselves)—and in both cases, the invaders are wreaking havoc. We would benefit from manipulating Hypersea's currents wisely."[60] Despite Zimmer's enthusiasm for Hypersea, the concept does not seem to make much of a difference, as the realization that humans are "stirring it up" merely echoes other familiar environmentalist warnings not to interfere with established ecological systems. The precautionary principle, by now quite established in environmental circles, does not require the Hypersea.[61]

Stefan Helmreich analyzes the genomic connections between terrestrial human bodies and sea life in his poetic and theoretically astute study *Alien Ocean: Anthropological Voyages in Microbial Seas* as well as in his article "Human Nature at Sea." In the introduction to *Alien Ocean* Helmreich writes, "Some readers may object that I have not written *Our Oceans, Ourselves*, a book that would highlight human intimacy with the sea, that would emphasize a sense of oceanic communion."[62] Although he admits that book is "indeed here, though submerged," he explains *Alien Ocean* is "skeptical of any simple identification with the sea, pessimistic about whether scientific knowledge alone about the ocean is enough for making sense of it (let alone protecting it)."[63] Helmreich does emphasize, however, the transformation of the human, explaining that the work of marine microbiologists makes it "possible to imagine elements of the human and the oceanic flowing into one another at a molecular scale. It allows scientists newly to describe human bodies as porous—to ocean-borne viruses and bacteria for example. It may become appropriate to think about the possibility that human nature, genetically understood, may be dissolving, a dissolution accomplished through the turbulent flowing together of human and oceanic biology."[64] This aqueous posthumanism, which is not unlike trans-corporeality in its insistence on the porousness of human bodies, challenges us to imagine how the "human," at the level of the gene, sloshes around with the rest of oceanic life. His conclusion notes the shift from recent accounts of the human as salty like the seas to the human as comprising bacteria:

> Once upon a time, the *human*, plunged into the sea (as blood, sweat, tears, milk), was baptized into communion with the planet. But plunged into the sea as a swirl of microbial genes, something unsettling happens. Microbes are not simple echoes of a left-behind origin for humans, orphaned from all evolutionary association. Microbes are historical and contemporary partners, part of our bodies' "microbiomes." "The" human genome is full of their stories, revealing that all genomes are metagenomes. The links between the scale of human bodies and ecologies become baroque, spatially and temporally. The bacteria that inhabit our bodies do not simply mirror the bacteria that inhabit the sea—as might brine in our blood. This is not human nature reflecting ocean nature. It is an entanglement of natures, an intimacy with the alien.[65]

Whereas accounts of how the human evolved from the sea, such as *Your Inner Fish*, trace a trajectory that culminates in a static and separate contemporary human, Helmreich elucidates how the human as part of the sea's "swirl of microbial genes" makes microbes our "partners," both historically and in the present. This sense of entanglement, which traverses realms, suggests ongoing material intra-actions as well as the inability to secure a human self as distinct from the "alien" other. This may be a potent posthumanism, which, as Helmreich concludes, saturates "human nature by other natures"[66] and reveals our complex and continuous entanglements with other lifeforms. I worry, however, along with one of the scientists Helmreich cites, that "we are losing sight of the organism."[67] While very little is known about the ecologies of marine habitats, because they are so difficult to study and because such studies often do not attract funding, Helmreich explains that gene sequencing, which is a perfect tool for "blue–green capitalism," allows marine microbiologists "to dispense entirely with the need to zero in on individual microbes—or even populations of discrete cells. . . . This is a genomics beyond organisms, a practice that implicitly queries whether individuals are the only evolutionarily meaningful units."[68] When so many marine animals are threatened with extinction, when so many marine ecosystems are on the verge of collapse, envisioning blue and green lifeworlds as one vast genomic soup may dramatize "our" corporeal intimacy with "the seas" but will not enable the sorts of engaged knowledges that trace how human practices threaten particular creatures, habitats, and ecologies. In my critique of *Cracking the Ocean Code*, a film that extols Craig Venter's massive project to mine the oceans for genetic material in order to translate "life to disk," "biochemical information" to "digital code," I argue that it would be more useful to consider marine life as always interconnected with particular environments, processes, and substances.[69] Such considerations would be highly mediated, but they would not culminate in the discovery of isolated genes, which become mere fodder for biocapitalist ventures. Instead, they would lead from "entanglement to greater entanglement" (with reference to Latour) as they trace interactions between forms of sea life, their environments, and the anthropogenic threats to species survival.[70]

Steve Mentz, in *At the Bottom of Shakespeare's Ocean*, warns that evocations of evolutionary connections to the sea may not be all that helpful: "Look at the world through salty eyeballs, remembering that the fluid in our eyes tastes like the sea. Most of the world is water. Most of that water is salt. No matter what it looks like, what it makes us feel, how our bodies float on its swells, the ocean is no place to live. . . . Long ago we crawled out of the water. We can't go back."[71] As Mentz suggests, nostalgia for our deep evolutionary past before tetrapods crawled up onto the land does not direct us toward solutions to current environmental predicaments. Furthermore, such origin stories, even as they attempt to elicit concern for the sea as our original home or for sea creatures as kin, revel in a prelapsarian innocence, as they skip a wide swath of history in which humans slaughtered ocean creatures and destroyed ocean ecologies, eliminating an astonishing number of marine mammals, birds, and fishes, by killing them for oil, fertilizer, fur, or food, or destroying their habitats through destructive trawling, the inadvertent production of dead zones, dumping, drilling, and development.[72] While Mentz contends that we need "more improvisational stories of working-with an intermittently hostile world,"[73] I think we need trans-corporeal modes of analysis that take responsibility for human actions within, and as part of the world. Barad stresses that the world is a "dynamic process of intra-activity," in which nothing exists that precedes relations. Emphasizing the dual meaning of "mattering," she defines ethics as "intra-acting from within and as part of the world in its becoming."[74] Trans-corporeality situates the (post)human as always already part of the world's intra-active agencies. For an oceanic sense of trans-corporeality to be an ethical mode of being, the material self must not be a finished, self-contained product of evolutionary genealogies but a site where the knowledges and practices of embodiment are undertaken as part of the world's becoming. Trans-corporeality, as part of risk society, requires that ordinary citizens have access to scientific information. But trans-corporeality as an ethical practice requires not only that citizens seek out information—which may or may not exist in any trustworthy or usable form—about risks to their own health but also that they seek out information about how their own bodily existence—their consumption of food, fuel, and specific consumer products—affects other people, other animals, habitats, and ecosystems. Tracing how terrestrial human bodies are intertwined with ocean ecologies is daunting, and yet, as I will discuss below, environmental art and activism are emerging that dramatize the material agencies that lurk in the most ordinary and seemingly benign objects and practices.

Radiation, Mercury, Plastic: New Materialism and Marine Activism

Shubin demonstrates how a recognition of the aquatic evolutionary origins of the human can coexist with conventional scientific epistemologies of distancing and disengagement, as well as with a medicalized sense of self-contained, genetically driven human bodies. Even as Rachel Carson's narrative evokes evolutionary kinship across vast temporal and oceanic expanses, thus working to dispel the sense that the seas are alien and separate from the human, such mythical stories, even when they conclude in present-day, palpable human bodies, may be dismissed as ancient history or immaterial myth. Oceanic origin stories can barely begin to matter if they do not open out to the ongoing material agencies of the present moment, acknowledging human culpabilities and vulnerabilities. Ten years after publishing

The Sea around Us Carson added a preface, contextualizing the book within the "atomic age," warning that the dumping of rubbish and radioactive waste will have catastrophic consequences for life itself. Carson explains that radioactive waste will be widely distributed not only by the water's movement, but also through living creatures, who, unknowingly, distribute radioactivity throughout the global seas: "The smaller organisms regularly make extensive vertical movements upward toward the surface of the sea at night, downward to great depths by day. And with them goes whatever radioactivity may be adhering to them or may have become incorporated into their bodies. The larger fauna, like fishes, seals, and whales, may migrate over enormous distances, again aiding in spreading and distributing the radioactive elements deposited at sea."[75] Although Carson states in *The Sea around Us*, first published in 1950, that "Man" "cannot control or change the ocean,"[76] a decade later she no longer imagines the ocean to be impervious to human harm. The mythical marine origin story magnifies the enormity of the threat of nuclear waste, rather than purifying it or dispersing it into oblivion. Carson does not herself trace the potential for radioactive waste dumped at sea to eventually enter human bodies, but since the next chapter outlines how we all carry the sea within ourselves, the reader may be given pause. Moreover, she leaves us with a disturbing recognition that containment is not possible, as animal bodies are not only permeable and vulnerable but also, through their usual movements and migrations, have become the distributors of dreadful anthropogenic threats.

Would that we now had Carson's words on the 2011 Fukushima nuclear disaster, which flooded the Pacific Ocean with radiation. The Center for Marine and Environmental Radiation at Woods Hole Oceanographic Institute launched a citizen science project to trace the movement of radioactive waters from Japan, across the Pacific, to the shores of North America. The organization's website explains the complex intra-activity between what we would ordinarily consider separate entities—water, organism, and sediment: "Scientists are tracking the many pathways by which radioisotopes from the damaged nuclear reactors at Fukushima make their way into and out of seawater, marine life, and seafloor sediments. These depend on the behavior and metabolism of individual animals, the nature of complex coastal and open-ocean processes, and the physical and chemical properties of individual isotopes."[77] Below this is the "'Tale of a Tuna,'" showing a map of a tuna contaminated with cesium-135 traversing the Pacific. The next panel, scrolling down, explains the effects of radiation on human health. The site represents the flows of radiation through water, marine life, and terrestrial human bodies. And even though the focus here is on the Fukushima disaster, it also contextualizes that catastrophic event with a fact that implicates the United States: "The primary source of cesium-137 has been nuclear weapons testing in the Pacific Ocean."[78] Strangely, however, Ken Buesseler, a scientist featured in a brief video on the site, downplays the dangers, restricting his concern to particular humans—to the Japanese people on land, who, of course, suffered immensely, and to the vague category of people who eat too much contaminated fish. His reassurance rings a bit disingenuous, however, given that dangerous levels of fish consumption are not demarcated.

BlueVoice.org, a marine conservation organization, epitomizes a trans-corporeal environmental politics by stressing that humans, dolphins, and whales are all vulnerable to the harmful effects of mercury and organochlorines. In the short film titled "A Shared Fate," Hardy Jones, a cofounder of BlueVoice.org, explains how he had dedicated his life to studying and protecting dolphins and whales. Ironically, his "extraordinary bond" with dolphins becomes undeniably corporeal. He explains, "I was diagnosed with a disease that intertwined my life with dolphins in a way I could never have imagined."[79] Jones had developed chronic mercury poisoning from eating the same fish that dolphins eat. He also developed multiple myeloma, which, as Dr. Brian Lurie explains, also affects dolphins, because dolphins "do not break down type II dioxins, and that puts them at risk, so we are now evaluating the same kinds of things in people."[80] Coal-burning power plants, pesticides, and flame retardants all result in an ocean riddled with mercury and organochlorides, which threaten marine life. Those who eat marine animals—for example, people who eat dolphins—suffer from high levels of dangerous heavy metals in their bodies. Ironically, it is the fact that dolphins and whales have become so toxic that may rescue them from slaughter. As with most transcorporeal recognitions in risk society, "A Shared Fate" displays both the necessity for scientifically derived data and the need for embedded epistemologies that reconfigure the boundaries between scientific practices, politics, human health, and environmentalism.

At the same time that the new materialisms are emerging across different theoretical domains, including the environmental human-ities, environmental movements and practices are emphasizing the un-settling and unintended consequences of substances and things. The Great Pacific Garbage Patch organization, for example, describes the magnitude of the Pacific Gyre, which is "roughly the size of Texas, con-taining approximately 3.5 million tons of trash. Shoes, toys, bags, paci-fiers, wrappers, toothbrushes, and bottles too numerous to count."[81] Everyday, ostensibly benign human stuff becomes nightmarish as it floats forever in the sea. The recognition that these banal objects, in-tended for momentary human use, pollute for eternity renders them surreally malevolent. Chris Jordan's stunning series of photographs, "Midway: Message from the Gyre," feature decomposing marine birds—the remnants of their carcasses revealing the pieces of plastic they have ingested.[82] These ghastly photographs display the painful contrast between the muted browns and grays of the decomposing bodies, bodies that are already becoming part of "nature" again, and the eerily cheery, super-colorful bits of plastic, predominantly bottle caps—the banal but persistent detritus of consumerism. Ironically, the birds, like good environmentalists, will "reuse" these bits of plastic, taking them from the site of the decomposed bodies and then eating them or feeding them to their young. One bottle cap—such a negli-gible bit of stuff to humans—may persist in killing birds and fish for hundreds (thousands?) of years. There is something uncanny about or-dinary human objects becoming the stuff of horror and destruction; these effects are magnified by the strange jumbling of scale in which a tiny bit of plastic can wreak havoc on the ecologies of the vast seas. We cannot see mercury or other chemicals within sea mammals, but these photographs disclose trans-corporeality—animal bodies invaded by terrestrial, human consumerism, revealing the swirling natural-cultural agencies, the connection between ordinary terrestrial life and ocean ecologies, and the uneven distribution of harm.

Jane Bennett in *Vibrant Matter: A Political Ecology of Things* contends that one of the reasons to "advocate the vitality of matter" is that "the image of dead or thoroughly instrumentalized matter feeds human hubris and our earth-destroying fantasies of conquest and consump-tion."[83] I agree. Grappling with what it means to understand mate-riality as agential, rather than as passive, inert, and malleable, is at the heart of new materialist theory. While particular strands of thing theory, object-oriented ontology, speculative realisms, new vitalisms, and material feminisms may or may not be particularly posthumanist or environmentally oriented, material ecocriticism, by definition, focuses on material agencies as part of a wider environmentalist ethos that values ecosystems, biodiversity, and nonhuman life. Serpil Oppermann, in "Ecocriticism's Theoretical Discontents," argues that we need to "advance a critical perspective in which both discursivity and materiality . . . can be integrated in a relational approach," and that the "accountability of such an approach must . . . lie in a correct identification of the ethical, epistemological, and ontological concerns of ecocriticism's wider interest in human and nonhuman systems."[84]

Attention to material agencies is not limited to academic scholarship, but instead is emerging across different domains, as environmental activists, movements, artists, and practices emphasize the unsettling and unintended consequences of substances and things. Indeed, there is a striking sort of "new materialisms" pulsing through green subcultures, as amateur environmental practitioners think through the strange agencies of ostensibly unremarkable substances, systems, and objects.[85] Climate change, sustainability, and antitoxin movements make environmentalism a practice that entails grappling with how one's own bodily existence is ontologically entangled with the well-being of both local and quite distant places, peoples, animals, and ecosystems. Campaigns against plastic link not only coastal regions but also inland zones to the mushrooming plastic found in the oceans. While plastics escape the ravages of time, a study on plastic pollution published in 1973 seems ancient as it concludes that plastic's harm is "chiefly aesthetic," since the "inert nature of plastic means that it is unlikely to enter the food chain and threaten human welfare."[86] Plastics are now known to absorb toxins, release toxins, and enter the marine food web. Greenpeace warns of plastic's "sinister twist": "The plastics can act as a sort of 'chemical sponge.' They can concentrate many of the most damaging of the pollutants found in the world's oceans: the persistent organic pollutants (POPs). So any animal eating these pieces of plastic debris will also be taking in highly toxic pollutants."[87] Plastics affect not only the larger and more visible sea creatures but also the very small, including those in the pelagic and deep seas. "Toxin-laden microplastics may add another risk to marine life," as the many creatures such as benthic worms, sea cucumbers, and krill "will ingest tiny plastic particles."[88] Ulrich Beck's risk society descends to the bottom of the sea, as the benthic creatures can no longer depend on their own sensory organs to detect danger. Their ways of knowing and being have been rendered inadequate by the xenobiotic substances that surround them. The anthropocene planet, littered with dangers that no species evolved to survive, overwhelms the ability of countless creatures to discern and adapt to threats.

In "Plastic Materialities" Gay Hawkins, drawing on Bennett's theory of "thing power" and vital materialism, which asserts "that things have the capacity to assert themselves," directs her attention to plastic bags: "As scientists discover marine life choking on bags and environmental activists document the bags' endless afterlife in landfills, plastic bags are transformed from innocuous, disposable containers to destructive matter."[89] Hawkins asks, concerned that these formulations do not pay sufficient attention to the bag itself, "But what of the bag in all this? It appears as a passive object of reclassification."[90] Charging that "ethics slides into moralism" when humans "are not invited to be open to the affective intensities of plastic matter; rather they are urged to enact their ethical will and eliminate it," Hawkins sets out to let "plastic bags have their say."[91] After analyzing say-no-to-plastic-bag campaigns, an everyday encounter with a sticky plastic bag, and the floating plastic bag made famous in the film *American Beauty*, she concludes by advocating Bill Connolly's conception of "critical responsiveness," which, she explains, "decenters the human as the sovereign source of agency and change; in recognizing multiple sites of agency at play in the world it invites an expanded politics attentive to how the force of matter might participate in generating new associations and ethics."[92]

As a new materialist I agree with Hawkins that "recognizing multiple sites of agency at play in the world invites an expanded politics"; indeed, trans-corporeality contends that the recognition of intra-active material agencies expands and transforms political and ethical domains. While Hawkins condemns the say-no campaigns because the "differential agency of the bag in this process is disavowed," as it "is something to be controlled by human will, not a participant in an emergent ethical constituency,"[93] I must confess that I cannot imagine how a plastic bag can be part of an "ethical constituency." While the surprising material agencies and effects of the plastic bags certainly exceed political and ethical frameworks—which is the key problem— that does not mean we should imagine bags as entities that contain their own voice, perspective, or rights. Although Hawkins's approach perceptively accounts for the thing power of the plastic bag, the bag is taken as a separate, discrete object rather than a phenomenon within larger economic, political, and environmental systems. By contrast, Barad insists we are responsible to others because of the "various ontological entanglements that materiality entails." Thus ethics is "not about right response to a radically exterior/ized other, but about responsibility and accountability for the lively relationalities of becoming of which we are a part."[94] We are always on the "hook"—on innumerable hooks—ethically speaking, always caught up in and responsible for material intra-actions. The theory of "intra-action" does not take separate, distinct, objects as a starting point, emphasizing instead that "relata do not preexist relations."[95] While Barad does not (and could not) offer specific guidance as to how to determine what particular ethical practices would entail, she emphasizes ontological entanglements rather than encounters with discrete objects. In a related fashion, my conception of trans-corporeality does not concentrate on bodies, things, and objects as separate entities,[96] but instead traces how the (post)human is always already part of intra-active networks and systems that are simultaneously material, discursive, economic, ecological, and biopolitical.

Whereas Hawkins distinguishes her approach to plastic bags from the moralism of environmental campaigns, which she claims disavow the bags' agency, I relish the parallels and affinities between new materialist theories and environmental activism. The very sense of ethics Barad describes, that of being responsible for the "lively relationalites of becoming of which we are a part," infuses campaigns that stress the unintended consequences and surprising material agencies of everyday objects. Jeffrey Jerome Cohen argues that the philosophy of "agentism" is itself a mode of activism that spurns anthropocentrism: "Agency is distributed among multifarious relations and not necessarily knowable in advance: actions that unfold along the grid surprise and then confound. This *agentism* is a form of activism: only in admitting that the inhuman is not ours to control, possesses desires and even will, can we apprehend the environment disanthropocentrically, in a teetering mode that renders human centrality a problem rather than a starting point."[97] Even while activist organizations target humans, attempting to change ideologies, beliefs, and behaviors, this trajectory does, in fact, "teeter" as human centrality is unsteadied by unruly nonhuman agencies. Activist organizations such as the Plastic Pollution Coalition, for example, devoted to "working towards a world free of plastic pollution and its toxic impacts,"[98] creatively demonstrate through videos and artworks the surreal and uncanny effects of the banal consumer objects that populate our world. Jonas Benarroch's brilliant two-minute video "The Ballad of the Plastic Bag" follows the plastic bag as it flies from a parking lot, across a prairie, over train tracks, above a scenic stone outcropping in the desert, over roads, fields, and forests, finally landing in a beautiful mountain lake where it sinks, just barely visible. The ironic, romantic lyrics, sung over a plucky guitar, are reminiscent of a drifter's ballad, a free-roaming spirit "traveling light": "But honey, I won't be chained / This spirit can't be tamed"; "Nobody puts a hold on me / And nothing can destroy my glee." The concluding caption is less cheery: "Plastic is not biodegradable. Its particles enter the food chain, intoxicating all organisms."[99] Although the film doesn't portray the agency of the bag as it releases toxins or clogs an animal's digestive track, the clever conceit of the plastic bag as a ramblin' man dramatizes the agency and "freedom" of this supposedly inanimate object, stressing that these flimsy things have gotten away from us, escaping human control. Rather than demonizing the object, the video invites the viewer to take vicarious pleasure in the bag's free-roaming, aesthetically pleasing travels. But the video is done in a playful and ironic spirit; we are not called to listen to the bag's perspective or extend our ethical concern toward it as an entity. It may provoke affection, but irony renders such feelings slippery. Although the film intends to convey an environmentalist message that plastic bags, in their freewheeling ways, cause serious harm, the pleasure provoked by the humorous song and visual narrative is not simply arrested or deflated by the concluding message that toxic plastic particles enter the food chain. Instead, the ironic pleasure the video provides suggests the daily practices of sustainable living that proceed from environmental movements are rich with passion, ingenuity, humor, and lively modes of critique. Notwithstanding the ominous warning that concludes the video, its billowing pleasures are akin to Rosi Braidotti's sense of a "non-rapacious ethics of sustainable becoming: for the hell of it and for the love of the world."[100]

Another video endorsed by the Plastic Pollution Coalition, a fifty-second spoof by Katrin Peters called "Plastic Seduction," features a romantic seafood dinner on a secluded beach. The man lovingly feeds the woman an oyster, and as she opens her mouth rather suggestively, we notice that the oyster sports a blue plastic bottle cap—which the woman crunches in delight (Figure 7).[101] They both act like nothing is awry as they dig into a plate of seafood mixed with colorful plastic garbage. While the couple's passion is not dampened by their bizarre meal (they exit, stage right, to indulge in other bodily pleasures), the camera pans out to reveal that the beach where they had been dining is full of plastic garbage, and the voice-over concludes, "Not so tempting after all. Help turn the tide."[102] The couple's dreamy mood suggests the power of plastic to seduce us all into a blissfully ignorant consumerism. While it is unlikely anyone will be served a plate of oysters topped with colorful plastic bottle caps, there is evidence to suggest that nearly all "seafood" humans consume has been contaminated by the staggering amount of plastics that have invaded the oceanic food webs. The voice-over explains, "Every year, thousands of tons of plastic ends up in our oceans. Plastic doesn't biodegrade in the sea. Over time it breaks up into tiny particles. Like sponges, these attract pollutants from the surrounding waters, accumulating a highly toxic chemical load before they contaminate the marine food chain."[103] "Plastic Seduction" dramatizes a trans-corporeality in which humans ultimately consume the surprisingly dangerous objects they have produced and discarded. The crunchy, colorful plastic pieces become metonyms of—not foils for—the actual seafood on the plate, which already harbors plastics and other toxins. While Ian Bogost, in *Alien Phenomenology*, wonders "what is it like to be a thing," such as the "udon noodle or the nuclear warhead,"[104] "Plastic Seduction" suggests something equally weird but more significant: ostensibly discrete entities such as plastic bottle caps are, in a sense, already part of who we are, as human diets ontologically entangle us with the plastic seas. Nancy Tuana, in "Viscous Porosity: Witnessing Katrina," describes how plastic not only invades human flesh but also exerts strange agencies once it takes up residence there. Describing herself breathing air polluted by plastic incineration, she states, "Components of the bottle have an agency that transforms the naturally occurring flesh of my body into a different material structure than what occurs in nature."[105]

Captain Charles Moore, who is known for discovering, researching, and publicizing the Great Pacific Garbage Patch, stresses the harmful agencies of seemingly benign objects. In his book *Plastic Ocean: How a Sea Captain's Chance Discovery Launched a Determined Quest to Save the Oceans*, Moore charges that the Plastic Age "has sneaked up on us almost imperceptibly," and that for awhile "we weren't as bothered

Figure 7. Still from the advertorial spot "Plastic Seduction" by Katrin Peters / Daily Difference–Media for Change.

as we might have been, because we still thought plastic material was inert and benign, an eyesore that couldn't do much harm."[106] He contests the assumption that plastic is inert by dramatizing the lively actions of trillions of seemingly inert objects. "Nurdles," for example, the preproduction pellets, "escape the distribution system and go feral, with billions eventually winding up in waterways and the oceans."[107] Ironically, the fact that the nurdles "go feral" invokes wildness, which has long been valuable for environmentalists because "it is not entirely dominated, monitored, transformed, and constrained or made to conform to the dictates of its efficient utilization by humans."[108] Plastic, arguably the quintessential substance for efficient domination, somehow manages to escape, mocking both the human mastery of the material world and the green ideal of wildness, as it multiplies and roams, garish and ghastly. Moore continues to animate and anthropomorphize this substance that surrounds us but ordinarily goes without notice: "Plastic is athletic. It scoots, flies, and swims. It travels without passport, crosses borders, and goes where it is, literally, an illegal alien. It has the endurance of a champ."[109] While the comparison to an illegal alien is terribly unfortunate, the overarching theme that Moore presents is of everyday objects gone wild, contradicting the presumption that human intentionality directs and confines material things:

"On land, it's soothing to think that all those bottles and wrappers, all that cheap plastic stuff we handle every day, winds up in a landfill, safely sequestered from polite society. But here in mid-ocean we're finding hordes of escapees . . . Try as we may to control it, to hide it, to manage it—it mocks us and goes where it doesn't belong."[110] While the pathetic fallacy of the garbage as intentionally taunting us hardly seems scientific, the way Moore animates plastic stuff not only underscores how harmful—if not malevolent—plastic can be, it struggles to convey a sense of material agency that will prove plastic is doing harm. Moore explains that when he first attempted to enlist experts in his quest to clean up the seas, he was surprised to be told that the mere presence of a mammoth amount of garbage in the ocean was not enough to provoke concern: "It can't only be about the ugliness or wrongness of plastic garbage in the remote ocean. . . . It's about giving credence to the sense that plastic is *doing* something out there, something very possibly unhealthy, something *harmful*."[111] So Moore sets out, as a citizen–scientist, to demonstrate what plastic is doing.

It is well-known that plastic bags look like jellyfish in the water, confusing turtles and other creatures, but Moore explains that nurdles resemble fish eggs, the food of many seabirds, and, more broadly, that plastic, this spectacularly multifarious substance, mimics many sorts of "edibles in the marine environment."[112] Their resemblance to food means the plastic bits, the plastic bags, the plastic objects beckon, entice, and deceive birds, turtles, fish, and sea mammals. Vivid examples of animals occupied by plastics reveal the unfortunate results: "Many of the salps we encounter sport plastics, inside and out, little colorful chips embedded in clear tissue."[113] Moore tells of one whale necropsy that revealed "nearly six square yards of compressed plastics, mostly shopping bags" which were taken from the animal's gut. Another uncovered "sweatpants, a golf ball, surgical gloves, small towels, plastic fragments, and twenty plastic bags."[114] And another whale was full of fishing debris (ghost nets), including a piece that was forty-five square feet.[115] He tells of the Malibu sea lion that had thirteen plastic shopping bags in its stomach. Dr. Lauren Palmer, the veterinarian at the Marine Mammal Care Center, thinks that "the neurotoxic effects of domoic acid might have spurred the sea lion to eat bags when normally she never would have."[116] Domoic acids are produced by harmful algal blooms, which may be triggered by sewage and fertilizer runoff from land. Captain Moore urges us to consider that plastics, far from being inert, benign objects, act in the ocean as if they were "predators": "Plastics could even be considered, in a sense, 'predators,' given the deadly nature of 'ghost fishing' and entanglements of marine turtles, mammals, pinnipeds, and cetaceans. Though plastic is not a living organism, it acts like one and has the impact of one and should be taken into account in characterizations of the ocean biome. What is most shameful in this more realistic modern scenario is that plastic, in a sense, is man's surrogate, swimming with the fishes and doing harm."[117] Moore's characterization of plastics as predators seeks to account for the many ways they kill ocean creatures and devastate marine ecosystems. Stressing that plastic is "man's surrogate" is a powerful rhetorical move, as we imagine plastic as a horrific extension of ourselves, a discarded and disavowed entity that bobs along, wreaking incalculable harm. As Serenella Iovino eloquently puts it, waste is "the other side of our presence in the world, our absence made visible."[118] Even where we are absent, plastic, Moore suggests, entangles us in ghastly nets of responsibility. Perhaps it is anthropocentric to imagine plastics as our surrogates, since this figuration diminishes the "feral" agencies of plastic and its ability to exceed human control. Nonetheless, the formulation captures how individual objects have surprising agencies, yet those agencies emerge from and act within wider economic, industrial, consumerist, and ecological systems of which we are always a part and for which industrialized humans bear responsibility. Plastics do not manufacture, purchase, distribute, or dispose of themselves.

Since 2006, Pam Longobardi, in her Drifters Project, has gathered plastic pollution from sites around the world, cleaning beaches and creating artworks, installations, exhibits, and community events. Her large wall installation Crime of Willful Neglect (for BP), made up of "429 pieces of vagrant oceanic plastic from Greece, Hawaii, Costa Rica, and the Gulf of Mexico," articulates plastic pollution with oil disasters (Figure 8). She explains her method as follows: "I approach the sites as a forensic scientist, examining and documenting the deposition as it lay, collecting and identifying the evidence of the crime."[119] The installation is artfully arranged, as the shapes parallel one another, forming lyrical visual patterns. The attention the artist devoted to the work affects the viewer, not least of which because the objects had been mere garbage, discarded things for which no one else had taken responsibility. Now they elicit a response. Crime of Willful Neglect (for BP) (2014) was part of the exhibit What Once Was Lost Must

Figure 8. Pam Longobardi, *Crime of Willful Neglect (for BP)*, 2014.
84 × 138 × 6 inches. 429 pieces of ocean plastic from Greece, Hawaii,
Costa Rica, and the Gulf of Mexico. Courtesy of Pam Longobardi;
copyright Pam Longobardi.

Now Be Found: Chronicling Crimes against Nature, a title that supplants the historically homophobic category of such "crimes" with
humanly made objects such as drift nets that harm sea life. While the
danger of evoking crimes against nature when displaying pollution
is that it could insinuate that homosexuality is unclean, the exhibit
itself, assembled with such care, suggests the discarded objects are
not made abject. The objects, neatly arranged and aesthetically striking, are nonetheless simultaneously the evidence and substance of
wrongdoing. Longobardi notes that "black plastic is the most ubiquitous and least recyclable type of plastic," indicating that the substance
itself is an ecological offense. But the color of the objects also suggests
birds and sea creatures that have been covered in oil. The shape of the
work—a drop—evokes a drop of oil, a drop of seawater, or a human
tear—a simultaneity that stings. Stephanie LeMenager in *Living Oil:
Petroleum Culture in the Living Century* writes that "the human body has
become, in wealthier parts of the world, a petroleum *natureculture*,"
as even the "self-identified environmentalists are driving cars, using
petroleum-based plastics, walking on asphalt, filling our teeth with
complex polymers, and otherwise living oil."[120] Indeed, within that
drop—the shape of oil, water, and human grief—are seemingly benign objects, made from oil and transported with oil, the stuff of normal life in the industrialized world. The few brightly covered bits add
a liveliness to the piece, but even that cheery, bright energy warns of
harmful material agencies, which exceed human justice and temporal reckoning: "The Deep Water Horizon Disaster is a crime that has
not seen full justice and whose future long term damage continues
to unfold."[121]

This chapter began by considering oceanic origin stories, moved on
to the strange agencies of marine pollution, and will conclude with a
futuristic vision of a flooded world. Marina Zurkow's mesmerizing,
gorgeous animated video "Slurb"[122] depicts humans, crabs, birds, rays,
jellies, giant dragonflies, and hybrid creatures—some with human bodies and dog heads who howl as they row along, one with a human body
and fishlike face. (See the still image from this video on the cover of
this book.) The description of the video, on the Streaming Museum
website, states that even though it was "inspired by fictions, like J. G.
Ballard's prescient 1962 novel, *Drowned World*," "nothing is fiction (although everything is hybrid)."[123] The many jellies in the water, for example, reference scientific predictions that the seas will be overrun with

gelatinous creatures, due to the extinction of other species and the fact that jellies can somehow survive polluted waters. The next chapter will discuss anthropocene seas and ocean futures. For our purposes here, however, it is important to point out that Zurkow paints the human and humanesque bodies aquamarine in color, visually connecting them with the water, suggesting that they are saturated with their aquatic environment. Not only are the humans themselves aquatic, but there is no escape from the water for either the viewer or the inhabitants: in the six-minute excerpt from this nearly eighteen-minute video, there is no dry land nor even any sky within the frame. The many creatures are crowded together, slowly moving through scenes that include the spires of submerged buildings, partially floating automobiles, piles of tires, dead trees, and other garbage. The steady, poignant music, the repetitive paddling of humans and their hybrids, and the energetic dance of the crabs depict quotidian practices of postapocalyptic survival, where the seas are full of nothing but jellyfish and the humans, with their blue flesh, can no longer be oblivious to their aquatic origins or their submersion within material worlds.

6
Your Shell on Acid
MATERIAL IMMERSION, ANTHROPOCENE DISSOLVES

◇◇

Who is the "anthro" of the "anthropocene"? In its ostensible universality, does the prefix suggest a subject position that anyone could inhabit? While the term "anthropocene" would seem to interpellate humans into a disorienting expanse of epochal species identity, some accounts of the anthropocene reinstall rather familiar versions of man. Feminist theory, long critical of "man," the disembodied, rational subject; and material feminisms, which stress inter- or intra-actions between humans and the wider physical world, provide alternatives to accounts that reiterate man as a bounded being endowed with unilateral agency. And while the geological origins of the term "anthropocene" have spawned stark terrestrial figurations of man and rock in which other life-forms and biological processes are strangely absent, the acidifying seas, the liquid index of the anthropocene, are disregarded, even as billions of tiny shelled creatures will meet their end in a catastrophic dissolve, reverberating through the food webs of the ocean. Thinking with these aquatic creatures provokes an "ecodelic,"[1] scale-shifting dis/identification, which insists that whatever the "anthro" of the "anthropocene" was, is, or will be, the anthropocene must be thought with the multitude of creatures that will not be reconstituted, will not be safely ensconced, but will, instead, dissolve.

Anthropocene Vision

As the "anthropocene" joins "climate change" and "sustainability" as a pivotal term in public environmental discourse, it may be useful to consider how the novel category becomes enlisted in all too familiar formulations, epistemologies, and defensive maneuvers— modes of knowing and being that are utterly incapable of adequately responding to the complexities of the anthropocene itself. As chapter 4 discussed, the invisible, unmarked, ostensibly perspectiveless perspective is common in visualizations of climate change as a global phenomenon. For example, the Group on Earth Observation's System of Systems depicted the earth as a blank slate for information, an empty stage awaiting scientific acts of creation.[2] Feminist theory, especially material feminisms and posthumanist feminisms, offer cautionary tales, counterpoints, and alternative figurations for thinking the anthropocene subject in immersive onto-epistemologies. Whereas a critical posthumanism contests the human as a conceptual apparatus that underwrites ordinary practices of exploitation, the concept of the anthropocene testifies that *Homo sapiens* has "achieved" an exceptional feat, that of epoch-making planetary alteration. Take the title of Will Stefan, Paul J. Crutzen, and John R. McNeill's article, "The Anthropocene: Are Humans Now Overwhelming the Great Forces of Nature?" which concludes that "humankind will remain a major geological force for many millennia, maybe millions of years, to come."[3] The hand-wringing confessions of human culpability appear coated with a veneer of species pride. To think of the human species as having had a colossal impact, an impact that will have been unthinkably vast in duration, on something we externalize as "the planet," removes us from the scene and ignores the extent to which human agencies are entangled with those of nonhuman creatures and inhuman substances and systems.

As the capitalist rapacity of the few and the subsistence needs of the many result, unintentionally, in the vast obliteration of ecosystems and the extinction of species, modes of acting within economic, technological, and environmental systems, such as quotidian acts of consumption, seem worlds apart from the aesthetically rendered scenes that deliver a spectacular view of manufactured geographies to spectators positioned outside the action. The epistemological position of the "God's-eye view" that Donna Haraway critiqued in "Situated Knowledges" dominates many of the theoretical, scientific, and artistic portrayals of the anthropocene. Ironically, at the very moment that the catastrophes of the anthropocene should make it clear that what used to be known as nature is never somewhere else (even the bottom of the sea has been altered by human practices), the "conquering gaze from nowhere," the "view of infinite vision," the "God trick" of an unmarked, disembodied perspective reasserts itself.[4] Yet the ostensibly infinite perspective excludes so much. Claire Colebrook in *Death of the Posthuman* argues that the "very eye that has opened up a world to the human species, has also allowed the human species to fold the world around its own, increasingly myopic, point of view."[5] Strangely, this humanist myopia may manifest as visual tropes that view the world at sanitized distances. And "the world" in these images is an eerily lifeless entity, devoid of other species, as if the sixth great extinction had already concluded.

Prevalent visual depictions of the anthropocene emphasize the colossal scale of anthropogenic impact by zooming out—up and away from the planet. Andrew Revkin's essay in the *New York Times*, "Confronting the Anthropocene," begins with a photo of a glowing spider-shaped blob of gold against darkness, with the following caption: "Donald R. Petitt, an astronaut, took this photograph of London while living in the International Space Station."[6] *National Geographic*'s story "Age of Man," written by Elizabeth Kolbert, begins with a rather dystopian aerial photo of Dubai, in which the vivid aqua waterway only highlights the otherwise utterly brown, bleak cityscape.[7] The *Encyclopedia of Earth* begins its entry on "anthropocene" with a cylindrical map (flat and rectangular), showing "the earth at night, demonstrating the global extent of human influence."[8] The blog *The Anthropocene Journal* sets out a stark, but at least nongendered, cluster of terms in its subtitle: "People. Rock. The Geology of Humanity."[9] Despite the subtitle "The Geology of Humanity," with its ambiguous "of," which could intermingle humanity and geology, the images shown on the "State of the Art" posting, for example, detach the spectator from the scene. Moreover, the blog's banner image features a globe, as if seen from space, showing North America lit up in yellow and blue capillary-like lights. Félix Pharand-Deschênes, listed as an "anthropologist and data visualizer," created this image as well as other similar images that appear on his *Globaïa* website.[10] Scrolling down his "Cartography of the Anthropocene" page, one encounters a series of globes, each with patterns formed by lines marking roads, cities, railways, transmissions lines, and underwater cables.[11] The patterns of bright blue or shimmering gold lines that span the planet demonstrate the expansiveness of human habitation, commerce, and transportation networks, marking human travel, transport, and activity against a solid background that obscures winds, tides, currents, and the travels of birds, cetaceans, or other creatures. Nonhuman agencies and trajectories are absent.

Where is the map showing the overlapping patterns of whale migrations with shipping and military routes? Or the sonic patterns of military and industrial noise as it reverberates through areas populated by cetaceans? Or established bird migration routes, many of which have been rendered inhospitable to avian life? The movements, the activities, the liveliness of all creatures except for the human vanish.[12] And, once again, in the dominant visual apparatus of the anthropocene, the viewer enjoys a comfortable position outside the systems depicted.[13] The already iconic images of the anthropocene ask nothing from the human spectator; they make no claim; they do not involve nor implore. The images make risk, harm, and suffering undetectable, as toxic and radioactive regions do not appear, nor do the movements of climate refugees. The geographies of the sixth great extinction are not evident. The perspective is predictable and reassuring, despite its claim to novelty and cataclysm.

David Thomas Smith's photography is introduced on the *Artstormer* site with an epigraph by A. Revkin, "We are entering an age that might someday be referred to as, say, the Anthropocene. After all, it is a geological age of our own making."[14] The singular human agency, as well as the possessive phrase "our own," is notable. What sort of subject could have produced a geological age? Betsy Wills introduces the photographs, which unlike the images of the globe depict merely a particularly processed portion of the earth, using highly mediated data: "Composited from thousands of digital files drawn from aerial views taken from internet satellite images, this work reflects upon the complex structures that make up the centres of global capitalism, transforming the aerial landscapes of sites associated with industries such as oil, precious metals, consumer culture information and excess. Thousands of seemingly insignificant coded pieces of information are sewn together like knots in a rug to reveal a grander spectacle."[15] These constructions are grand spectacles indeed. The swirling baroque designs captivate. They urge viewers to shift scales and recognize how small alterations of the landscape may be multiplied into geographical immensity (Figure 9). This immensity, however, is safely viewed from a rather transcendent, incorporeal perspective, not from a creaturely immersion in the world. Moreover, although trees are visible, for the most part these landscapes are devoid of life; they depict hard, flat surfaces, planetary puzzle pieces. The aesthetic is one of order and symmetry within complexity, suggesting the possibility of and desire for exquisite, intricate manipulations. Despite the scaling up these are, to contradict Mina Loy, tame things despite their immensity,[16] as the

Figure 9. David Thomas Smith, 1000 Chrysler Drive, Auburn Hills, Michigan, United States, 2009–2010. Courtesy of David Thomas Smith; copyright David Thomas Smith.

world is rendered into a kaleidoscopic vision you may hold in your mind like a toy in your hand. The super-symmetrical structure of Smith's photos, however, with double mirror images, in which everything in the top half is repeated in the bottom half and everything on the left is repeated on the right, presents an implicit critique of the scale of human transformation of the earth, by dramatizing a claustrophobic enclosure in a world that, in its predictable repetitions, becomes all too human, all too structured. Smith's work encapsulates the problematic of the anthropocene, as its aesthetic seduces with its precise symmetries and the prospect of mastery, but ultimately confines the viewer in a place devoid of surprises. Brilliantly, its aesthetic pleasures are the selfsame as its critique, as its visual delights repeat in solipsistic symmetries. It may be fitting to invite Patricia Johanson, the environmental artist from chapter 1, back into the discussion here: "I believe human beings are increasingly threatened and impoverished by the relentless conversion of every scrap of territory for our own limited and temporary uses."[17]

Abstract Force

The concept of the anthropocene, with its geological reference and its undifferentiated "anthro," retreats to a simple equation of "man" and "rock," an oddly stark rendition when one considers that current biophysical realities can only be approached through scientific captures of a multitude of intersecting biological and chemical, as well geological, transformations, which intermesh human and natural histories. Even though the concept of the anthropocene muddles the opposition between nature and culture, the focus on geology, rather than, say, chemistry or biology, may segregate the human from the anthropogenic alterations of the planet, by focusing on an externalized and inhuman sense of materiality.[18] Dipesh Chakrabarty's momentous essay "The Climate of History" raises essential questions about the nature of the human, some of which, in my view, turn on the conception of species identity, corporeality, and agency. Chakrabarty's first thesis in this essay is "Anthropogenic Explanations of Climate Change Spell the Collapse of the Age-Old Humanist Distinction between Natural History and Human History."[19] Despite the collapse of distinctions, Chakrabarty brackets humans as biological creatures—our own corporeality as living beings becomes eclipsed by the enormity of our collective geological alterations. He writes, "Human beings are biological agents, both collectively and as individuals. They have always been so. There was no point in human history when humans were not biological agents. But we can become geological agents only historically and collectively, that is when we have reached numbers and invented technologies that are on a scale large enough to have an impact on the planet itself."[20] While we could read the phrase "biological agents" as meaning that humans *are* biological *and* act on the biological, the phrase "geological agents," which follows, delimits the first phrase to imply that humans have had an effect on biological entities—not that we are ourselves interwoven into living and nonliving trans-corporeal networks. Moreover, the distinction between biological and geological agency is not tenable, since biological and chemical transformations flow through the world in multiple and messy ways. And, of course, the origin of so many anthropocene alterations—the colossal output of carbon—is a matter of chemistry and, in epochal timescales, biology, as fossil fuels issue from decomposed organisms. The essay "The New World of the Anthropocene," published in *Environmental Science and Technology* by Jan Zalasiewicz and colleagues, states that "far more profound" than the "plainly visible effects . . . on the landscape" "are the chemical and biological effects of global human activity," including the rise of CO_2 levels, the sea level rise, the acidification of the oceans, and the sixth great extinction.[21] Attending solely to the lithic imports delusions of separation and control that have no place in the global biological, chemical, and geophysical intra-actions of the anthropocene. Yet Chakrabarty subordinates "man's" interactions with "nature," to the new paradigm in which humans become a geological force when he asserts, "For it is no longer a question simply of man having an interactive relation with nature. This humans have always had, or at least that is how man has been imagined in a large part of what is generally called the Western tradition. Now it is being claimed that humans are a force of nature in the geological sense."[22] While the idea that humans have become a "force of nature in the geological sense" may seem to merge humans with something called "nature," the abstract formulation of the "force" reinstalls "man" as a disembodied potency, outside the nature he would alter. Thinking human as "force" represents a retreat from the radical risk, uncertainty, and vulnerability of the flesh, as humans are rendered strangely immaterial. This immateriality, then, also creates an impasse for thinking in terms of species identity.

Chakrabarty's fourth thesis results in an impasse: "The Cross-Hatching of Species History and the History of Capital Is a Process of Probing the Limits of Historical Understanding."[23] Drawing on Gadamer, Chakrabarty contrasts "historical consciousness" as a "mode of self-knowledge" with what he claims would be an impossible achievement, "self-understanding as a species":

> Who is the we? We humans never experience ourselves
> as a species. We can only intellectually comprehend or
> infer the existence of the human species but never expe-
> rience it as such. There could be no phenomenology of
> us as a species. Even if we were to emotionally identify
> with a word like *mankind*, we would not know what
> being a species is, for, in species history, humans are only
> an instance of the concept species as indeed would be
> any other life form. But one never experiences being
> a concept.[24]

I would like to address this question rather indirectly, by shifting from Gadamer and broadening the framework to include a range of theories and perspectives on species being. While the question of "who is the we" is always at play, and will become more complicated below, to say humans have never experienced themselves as a species seems mistaken. It is hard to imagine that indigenous peoples would not have elaborated, within their cultures and traditional ecological knowledges, a sense of what it is to be human within a multispecies world. Elizabeth DeLoughrey in "Ordinary Futures: Interspecies Worldings in the Anthropocene" draws on Maori models of epistemology, for example, to offer an "alternative mode of understanding climate change than Dipesh Chakrabarty's argument that our awareness of ourselves as geological agents cannot be understood ontologically." In the Maori mode that she describes, the subject is incorporated "into planetary networks of kinship" in which "knowing and being are constitutive and interrelated."[25] In the West, Darwin's *Descent of Man* intensified a species consciousness even as it intermingled the human with other creatures as progenitors and kin. Even those who deny evolution proclaim a particularly *human* exceptionalism, which could itself be understood as a form of species identification, albeit with religious rather than scientific origins. Furthermore, contemporary environmental discourses address humans as one species among other species, seeking to ignite an ethical or political sense of being part of a community of descent that is only intensified by the recognition of human culpability so readily available in the anthropocene. More quotidian relations with other species could also be said to characterize phenomenologies embroidered with species recognition. Species is certainly a concept, but it is a concept that is as substantial and as close at hand as one's own morphology. One does not need to read Darwin to notice the ways in which one's body is similar to and different from that of other living creatures. Natural history museums, zoos, television programs, or face-to-face encounters with wild or domestic animals spark a sense of species identity that is not singular, but is generated from a sense of species in relation. Exhibit A may well be that of people comparing their own hands to the fins of whale or dolphin skeletons displayed at a natural history museum—kinship inscribed in the bones. Donna Haraway's work, from *Primate Visions* to *The Companion Species Manifesto* to *When Species Meet*, attests to multiple modes of cross-species encounters, relationships, and phenomenologies that can be understood as modes of species consciousness, in which humans are both embodied creatures dwelling in their own present moments and creatures able to imagine vast historical narratives such as the coevolution of humans and canines. As Haraway states, "The temporalities of companion species comprehend all the possibilities activated in becoming with, including the heterogeneous scales of evolutionary time for everybody but also the many other rhythms of conjoined process."[26]

Chakrabarty's assertion that no one ever "experiences being a concept" is also strange, given the body of scholarship focusing on how those who inhabit marked identities and subjectivities, those who have been cast outside the Western conception of "man" or "the human," have negotiated, resisted, and transformed identity categories and subject positions. Feminist theory, postcolonial theory, critical race studies, and cultural studies offer numerous accounts of the relation between subjects, identity categories, and other concepts such as "woman," for example, from Monique Wittig's claim that lesbians are not women because woman is a structural relation to man, to Gayatri Spivak's notion of strategic essentialism. The vertiginous intellectual work required to "be a concept" is evidenced by W. E. B. Du Bois's theory of "double consciousness," Frantz Fanon and Homi Bhabha's conceptions of mimicry, the feminist practice of "consciousness-raising," and Judith Butler's notion of "gender trouble."[27] A Lacanian theorist may contend that one always experiences oneself as something akin to a concept, in that the mirror stage testifies to the fundamental misrecognition of self as coherent whole, despite gaps and contradictions. These are, for the most part, politicized modes of knowing and being, not "pure" or abstract species consciousness, to be sure. Rory Rowan puts it quite well: "Anthropos can be understood not as a pre-constituted identity but rather as the object of political contestation in the struggle to define the terms of future human existence on the planet."[28] Rowan's sense of the "anthropos" as concept within the terrain of political struggle places the term where it belongs, in the messy space where science, history, cultural identities, and politics coincide. Ultimately, whatever it may mean to think oneself as a species will be inextricably bound up with other more local identities and cultural conceptions, rather than separate from them. The anthropos, despite the predominant visualizations that obscure local contexts, could provoke a sense of species identity quite different from the lofty Western, capitalist humanism, with the recognition that every member of the species is at once part of long evolutionary processes, a member of a species that has had a staggering impact on the planet, and an inhabitant of a particular geographic, social, economic, and political matrix, with attendant and differential environmental vulnerabilities, culpabilities, and responsibilities.

One of Chakrabarty's most significant provocations is that thinking the human species as geophysical force—more on that below—precludes attention to social justice. Ian Baucom notes Chakrabarty's "quite stunning turn to the concept of species; to a new thinking of freedom for human life in its biological totality; to a mode of universalism apparently antithetical both to his preceding philosophy of history; and to what Gayatri Spivak has called the practice of postcolonial reason." He adds, "Confronted with the arriving and coming catastrophes of climate change, freedom can no longer be conceived of as the freedom of difference against the power of the globalizing same."[29] Baucom captures the crux of the matter here, as the enormity of global environmental crises would seem to call for human collectivity that trumps all other differences. Jamie Lorimer notes that as a "growing body of critical work makes clear, scientific invocations of a planet-shaping Anthropos summon forth a responsible species—or at least an aggregation of its male representatives. A common 'us' legitimates a biopolitics that masks differential human responsibilities for and exposures to planetary change."[30]

This should give us pause, especially since scientific discourse gains legitimacy precisely through its free-floating "objectivity." Scientific neutrality lends itself to a mode of popularization that cleanses the term "anthropocene" from any entanglement with political genealogies, specificities, and identities. Indeed, the visual depictions of the anthropocene discussed above do just that by scaling up so that human poverty, drought, flooding, or displacement is obscured from sight and the viewer is not implicated, nor is someone potentially affected by climate disasters or slow violence.[31] Sylvia Wynter's work, although too complex to be adequately discussed here, is nonetheless invaluable for this debate. In the discussion between Wynter and Katherine McKittrick, titled "Unparalleled Catastrophe for Our Species? Or to Give Humanness a Different Future: Conversations," Wynter states, commenting not on Chakrabarty's question about who the "we" is but, instead, on Jacques Derrida's 1968 talk "The Ends of Man," which concluded with the same question:

> The *referent-we* of man and of its ends, he implies, *is not* the *referent-we* of the human species itself. Yet, he says, French philosophers have assumed that, as middle-class philosophers, their *referent-we* (that of Man2) is isomorphic with the *referent-we* in the *horizon of humanity*. I am saying here that the above is the single issue with which global warming and climate instability now confronts us and that we have to replace the ends of the *referent-we* of liberal monohumanist Man2 with the ecumenically human ends of the *referent-we in the horizon of humanity*.[32]

Wynter contends that to deal with climate change requires "a far-reaching transformation of knowledge," which includes the very definition of the human as such,[33] which she herself offers throughout her dazzlingly original theoretical work. Alexander G. Weheliye states, "Wynter's large-scale intellectual project, which she has been pursuing in one form or another for the last thirty years, disentangles Man from the human in order to use the space of subjects placed beyond the grasp of this domain as a vital point from which to invent hitherto unavailable genres of the human."[34] Wynter's project, disentangling man from the human, may address the quandary of the anthropocene in that it suggests that multiple "genres" of the human may be inhabited, which means that the term "anthropocene" does not require a new sort of univocal "man." Environmentalisms, movements for environmental justice, climate justice, social and economic justice, along with struggles for indigenous sovereignty, will no doubt emerge from particular, local formulations of the human, which may or may not be linked with the "anthropocene." In *Friction: An Ethnography of Global Connection* Anna Lowenhaupt Tsing argues that "universals are effective within particular historical conjunctures that give them content and force. We might specify this conjunctural feature of universals in practice by speaking of engagement. Engaged universals travel across difference and are charged and changed by their travels. Through friction, universals become practically effective."[35] As an engaged universal, the species identity of the anthropocene would not be free-floating, but instead conjunctural. How will the "anthropocene" travel and what sort of friction will those travels entail? Will the politically forged and conjuncturally specific conception of the anthropos enable new modes of struggle for social justice, environmental justice, climate justice, biodiversity, and environmentalisms?

One of the most intriguing concerns that Chakrabarty puts forth is the idea that the anthropocene means reckoning with humans as a "force." Some of his concerns, I would suggest, could be addressed by a more material conception of the human and a less unilateral sense of agency. He writes:

> But if we, collectively, have also become a geophysical
> force, then we also have a collective mode of existence
> that is justice-blind. Call that mode of being a "species"
> or something else, but it has no ontology, it is beyond
> biology, and it acts as a limit to what we also are in the
> ontological mode. This is why the need arises to view
> the human simultaneously on contradictory regis-
> ters: as a geophysical force and as a political agent, as
> a bearer of rights and as author of actions; subject to
> both the stochastic forces of nature (being itself one

such force collectively) and open to the contingency
of individual human experience; belonging at once
to differently-scaled histories of the planet, of life and
species, and of human societies.[36]

The shift from the abstract "geophysical force" to "species" is jarring,
given that species—a biological category—is said to have "no ontology" and to exist "beyond biology." I agree that the human must be
apprehended "simultaneously on contradictory registers" and scales;
indeed, this is something that my conception of trans-corporeality,
which is grounded in environmental justice and environmental health
movements, seeks to do. And as Rowan suggests, stressing the anthropos as an object of political contestation, rather than as an already
fossilized term, allows for differentiation of particular groups of humans, along the lines of culpability and exploitation, distinguishing,
say, indigenous Amazonian peoples whose lands have been destroyed
by oil companies from those who benefit from oil company revenues,
or middle-class U.S. citizens driving automobiles from the citizens
of Pacific islands being driven from their homes by rising sea levels.
Thinking the human as a species does not preclude analysis and critique of economic systems, environmental devastation and social injustice. In fact, if we shift from the sense of humans as an abstract
force that acts but is not acted on, to a trans-corporeal conception of
the human as that which is always generated through and entangled
in differing scales and sorts of biological, technological, economic,
social, political, and other systems, then that sort of human—always
material, always the stuff of the world—becomes the site for social
justice and environmental praxis.

In "Brute Force" Chakrabarty writes, "But to say that humans have
become a 'geophysical force' on this planet is to get out of the subject/
object dichotomy altogether. A force is neither a subject nor an object.
It is simply the capacity to do things."[37] Feminist theory, science studies, and environmental theory have long critiqued the subject/object
dualism, often by underscoring the strange agencies of the entities
considered inert objects. New materialisms emphasize materiality as
agential, stressing the entanglements and interactions between humans and the nonhuman world. Interactive material agencies may be
dispersed and nearly impossible to trace, delimit, or scientifically capture, but that does not mean they evaporate. Claiming that a force is

neither subject nor object, however, seems to dematerialize said force when, in fact, the anthropocene results from innumerable human activities, activities that humans have engaged in as ordinary embodied creatures and as rapacious capitalists and colonialists. The force is not as abstract as it would seem, since the activities, the processes, and the results are not at all immaterial, and not at all mysterious. Humans are not gravity.[38] Perhaps the term "force" leads us astray. Chakrabarty notes, "A force is the capacity to move things. It is pure, nonontological agency."[39] Just because the scale of humans as a "geological force" is so immense, nearly unthinkable from the minuscule moments of everyday life, does not mean that it is an entirely different entity. It is a matter of scale, not a difference of kind. Human beings, who eat, who heat and cool their homes, who plug in their electronic devices, who transport themselves and their goods, who use fossil fuels in their everyday lives, and who may or may not reckon with an environmental consciousness, are, ultimately, part of this supposedly "nonontological agency." Moreover, other accounts of the anthropocene, such as that of Zalasiewicz and colleagues, cited above, stress its biological and chemical dimensions—which are even more difficult to conceive as an abstract or pure force, apart from the messy interactions of material beings and the stuff of the world.

The anthropocene suggests that agency must be rethought in terms of interconnected entanglements rather than as a unilateral "authoring" of actions. Jessi Lehman and Sara Nelson argue in "After the Anthropocene," for example, that "humanity's agency as a geological force confronts us not as a product of our supposedly unique capacity as humans for intentional action (as described by Marx, 1867, in his comparison of the architect and the bee), but as an unintended consequence of our entanglements with myriad non-human forces—chief among them fossil fuels. The Anthropocene therefore simultaneously expands and radically undermines conventional notions of agency and intentionality."[40] Similarly, Derek Woods in "Scale Critique for the Anthropocene" contends that "assemblage theory is necessary to move beyond the notion that the 'species' is a geologic force," proposing that the "scale-critical subject of the Anthropocene is not 'our species' but the sum of terraforming assemblages composed of humans, nonhuman species, and technics."[41] Woods's argument is convincing, especially in that it addresses one of the ironies or paradoxes of the anthropocene: "The present is a moment of human disempowerment in relation to terraforming assemblages."[42] That is certainly the case, as processes have been set in motion that will have devastating effects for thousands of years. And yet, in the face of this shattering disempowerment, some groups of humans will, nonetheless, persist in attempting to do something. Modes of thinking, being, and acting may arise from a political recognition of being immersed in the material world, as they contend with the conceptual challenges of shifting timescales and traversing geo-capitalist expanses where one's own small domain of activity is inextricably bound up with networks of harm, risk, survival, injustice, and exploitation. Some activist practices, such as personal carbon footprint analysis and other "micro-practices of everyday life,"[43] already exemplify the attempt to understand the human as a geophysical "force," through politicized modes of knowing and acting that are immersed and contingent rather than disembodied.

Immersed, Enmeshed Subjects

To counter the dominant figurations of the anthropocene, which abstract the human from the material realm and obscure differentials of responsibility and harm, I propose that we think the anthropocene subject as immersed and enmeshed in the world. In contrast to *Globaïa's* "Cartography of the Anthropocene" maps discussed above, Nicole Starosielski's multimedia project "Surfacing," for example, portrays undersea fiber-optic cable systems in such a way that "the user becomes the signal and traverses the network."[44] The user is immersed in technologies, marine spaces, geographies, landscapes, and histories: "You begin on the coast, carried ashore by undersea cable. From your landing point, you can traverse the Pacific Ocean by hopping between network nodes. You might surface at cable stations where signal traffic is monitored, on remote islands that were once network hubs, and aboard giant ships that lay submarine systems."[45] The design deliberately frustrates attempts to gain a bird's-eye view or to escape, as the user is always positioned, always inside the system. And many of the photographs of particular places where the user surfaces, such as Vung Tau, Vietnam, or Papenoo, Tahiti, reflect a human scale, the ordinary perspective of a person with a camera. The photo of Pacific City, Oregon, places the viewer behind a worker operating heavy machinery and only slightly above the muck of the drilling site. While this beautifully designed project is not explicitly

about the anthropocene, it nonetheless encourages its users to experience the sort of built, global systems that have become emblematic
of the anthropocene—but in an immersed, never omniscient position.
The project does not simply scale up into representations that afford
transcendence, but instead demands scale shifting and imaginative
encounters with human and nonhuman agencies. Similarly, describing her book, Starosielski writes, "Rather than envisioning undersea
cable systems as a set of vectors that overcome space, *The Undersea
Network* places our networks undersea: it locates them in this complex
set of circulatory practices, charting their interconnections with a dynamic and fluid external environment."[46] By doing so, it offers "what
might be an unfamiliar view of global network infrastructure," which
brings "geographies back into the picture," and reintroduces, perhaps,
an "environmental consciousness, to the study of digital systems."[47]

The immersed subject of trans-corporeality reckons with the anthropocene as an intermingling of biological, chemical, and climatic
processes, which are certainly neither simply "natural" nor managed
by human intention.[48] The trans-corporeal subject emerges from environmental health and environmental justice movements, including the
citizen–scientists who must discern, track, and negotiate the unruly
substances that move across bodies and places. Thinking the subject
as a material being, subject to the agencies of the compromised, entangled world, enacts an environmental posthumanism, insisting that
what we are as bodies and minds is inextricably interlinked with the
circulating substances, materialities, and forces. Rhonda Zwillinger's
photographic volume *The Dispossessed* could be read as an example of
trans-corporeal inhabitations of the anthropocene.[49] Zwillinger documents how people with multiple chemical sensitivity (MCS) attempt
to fashion less toxic living spaces, portraying the human as coextensive
with the built landscape of consumerism, where everyday objects, the
domestic, and the desert landscape become scrambled and menacing.
In one photo a woman sits under her carport, surrounded by the stuff
that should be within a home—her bed, computer, and so forth—
cluttering the space which is neither indoors nor outdoors, but a hybrid
zone. This stuff, the ordinary things of late twentieth-century human
habitats, has unexpected, injurious agencies for those with multiple
chemical sensitivity; they penetrate the person, harming physical and
mental health. Zwillinger's photographs offer an intimate, tangible,
and everyday—rather than philosophically abstract—sense of anthro-

pocene scale shifting, as they ask us to imagine the domestic as linked to toxic networks of industrial production, consumer use, and disposal. They call the viewer to trace the invisible, interactive material agencies that cross through bodies and places, rather than removing the human from the scene. Set in the vast desert landscape, the makeshift and often confining living arrangements of those with MCS radiate outward in all directions, linking human homes to undomesticated but nonetheless contaminated landscapes. Zwillinger depicts the toxic anthropocene as unnervingly commonplace.

In Colebrook's brilliant and disturbing essay "Not Symbiosis, Not Now: Why Anthropogenic Climate Change Is Not Really Human," she contends, "The figural and critical truth of the Anthropocene is that just as there is no pure earth than might be reclaimed, so there is no thought that is not already contaminated and made possible by the very logic of man that ecology might seek to overcome."[50] Specifically, Colebrook points to recent theoretical turns that coincide with material feminisms and feminist posthumanisms, "these turns 'back' to bodies, matters, historicity, ecology and the lived," calling them "reaction formations or last gasps."[51] She asks, "What if all the current counter-Cartesian, post-Cartesian or anti-Cartesian figures of living systems (along with a living order that is one interconnected and complex mesh) were a way of avoiding the extent to which man is a theoretical animal, a myopically and malevolently self-enclosed machine whose world he will always view as present for his own edification?"[52] Since my conception of "trans-corporeality" qualifies as an anti-Cartesian figure of "living systems," as a "complex mesh," Colebrook's contention stings. And yet I wonder whether, as a feminist theorist, her use of "man" here intentionally allows for the possibility that feminist theories may somehow depart from the modes of thought produced by man as a "myopically and malevolently self-enclosed machine," even as they function within "already contaminated" thought. Although this is not the sort of contamination she had in mind, I would pose the trans-corporeal onto-epistemologies of those with multiple chemical sensitivity as an alternative to the self-enclosed theories of the world, as people with MCS register material agencies of substances that can never be imagined as external and that demand both an experiential and theoretical grappling with the precise ways in which self and world are intermeshed.

While it is one thing to conceptualize how toxins circulate through bodies and environments, it is another for humanities scholars and artists to conceptualize humans as enmeshed with something as rigid as a rock. Some scholarly and artistic engagements with the geologic shift scales in ways that are intimate and generative. In *Stone: An Ecology of the Inhuman*, for example, Jeffrey J. Cohen writes, "This book is something of a thought experiment, attempting to discern in the most mundane of substances a liveliness. Despite relegation to a trope for the cold, the indifferent and the inert, stone discloses a queer vivacity, a perilous tender of mineral amity." Cohen posits a "human–lithic enmeshment" as he analyzes the ecomaterialisms of the Middle Ages and contemporary theory, noting that stones "erode the boundary that keeps biological and mineral realms discrete."[53] The editors of the beautiful collection *Making the Geologic Now: Responses to Material Conditions of Contemporary Life* define their concept of the "geologic turn" with reference to practices that involve "exposure and visceral response to actual event-ness, or to change or forces."[54] Making "a geologic turn," they say, entails recalibrating "infrastructures, communities, and imaginations to a new scale—the scale of deep time, force, and materiality. . . . We do not simply observe [the geologic] as landscape or panorama. We inhabit the geologic."[55] And the geologic inhabits us. Ilana Halperin, an artist described as having "deep love of geology," writes in her essay "Autobiographical Trace Fossils" how the "boundary between the biological and geological can begin to blur."[56] Referring to "body stones," such as gall and kidney stones, she states, "In the body, each stone is a biological entity, and once out of the body it belongs to the realm of geology."[57] Kathryn Yusoff argues for a " 'geological turn' that takes seriously not just our biological (or biopolitical) life, but our geological (or geopolitical) life as crucial to modes of subjectification in the Anthropocene." She investigates what she terms "geologic life," "a mineralogical dimension of human composition that remains currently undertheorized in social thought."[58] Stephanie LeMenager in *Living Oil: Petroleum Culture in the American Century* explores museums, photography, literature, and other cultural productions as she documents an immersed, intimate, and unsanitized sense of dwelling in the anthropocene: "We experience ourselves, as moderns and most especially as modern Americans, every day in oil, living within oil, breathing it and registering it with our senses. The relationship is, without question, ultradeep."[59] As different as these projects are, none of them extracts the human from the world, but instead conceptualizes the human as intermingled with the lithic and the inhuman—the energy, matter, and temporalities of the geologic.

Your Shell on Acid: Anthropocene Dissolves

Notwithstanding the lively and generative thinking with stones, geologic life, and petrocultures, by Cohen, Yusoff, LeMenager, and others, I would like to contribute another sort of figuration of the anthropocene, which is aquatic rather than terrestrial. It is vital to contemplate the anthropocene seas, not only because marine ecosystems are gravely imperiled but also because the synchronic depth and breadth of the oceans present a kind of incomprehensible immensity that parallels the diachronic scale of anthropogenic effects.[60] The deep seas, once thought to house "living fossils" that terrestrial time left behind, are in fact home to sea creatures who live at a slower pace, within the cold, dark, and heavy waters. Oceanic depths, especially, resist the sort of flat mapping of the globe that assumes a "God's-eye view." The view of the earth from space reveals merely the surface of the seas, a vast horizontal expanse that is rendered utterly negligible when one considers the unfathomable depths and three-dimensional volume of the rest of the ocean. To begin to glimpse the seas, one must descend, rather than transcend,[61] be immersed in highly mediated environments that suggest the entanglements of knowledge, science, economics, and power. While the human alterations of the geophysical landmasses of the planet can be portrayed as a spectacle, the warming and acidifying oceans, like the atmospheric levels of CO_2, cannot be directly portrayed in images but must be scientifically captured and creatively depicted. The depths of the ocean resist flat terrestrial maps that position humans as disengaged spectators. Marine scientists must, through modes of mediation, become submerged, even as persistent Western models of objectivity and mastery pull in the opposite direction.[62] The substance of the water itself insists on submersion, not separation. Even in the sunlit, clear, shallow waters that divers explore, visibility is never taken for granted, nor does distance grant optimal vision. The oceans proffer a sense of the planet as a place where multiple species live as part of their material environs. As human activities change the chemical composition, the temperature, and the alkalinity of the waters, marine creatures also change.

Lesley Evans Ogden in "Marine Life on Acid," published in *BioScience* in 2013, explains that the term "ocean acidification" was coined only in 2003, yet this problem has already become known as "climate change's evil twin."[63] She explains what is happening:

> The ocean is a massive carbon sink estimated to have absorbed one-third of all the CO_2 produced by human activities. The tracking of carbon concentrations in the ocean, which began in the mid-1980s, has indicated that concentrations of CO_2 are increasing in parallel with the growing amount of this gas in the atmosphere. Short-term and long-term cycles continually exchange carbon among the atmosphere, the ocean, and land. CO_2 reacts with seawater to form carbonic acid, but as a weak acid, carbonic acid almost immediately dissociates to form bicarbonate ions and hydrogen ions. The increasing concentration of hydrogen ions makes seawater more acidic.[64]

Ogden notes that the ocean is "now nearly 30 percent more acidic than it was at the beginning of the industrial era" and that finding "a comparable acidification event" entails "going back 55 million years."[65] Research on how the shift in alkalinity affects sea life and ecosystems is only just beginning, but already a strange array of effects have been captured. Acidification makes the eggs of the red sea urchin not as quick as they need to be in blocking out a second sperm, resulting in inviable embryos; the tiny plankton *Ostreococcus tauri*, which is normally 1 micrometer enlarges to 1.5 micrometers with increased CO_2, which means that some creatures dependent on it for food may no longer be able to eat it.[66] The alteration of ocean alkalinity even causes confusion and destructive behavior in fishes—which is fascinating in its scrambling of biosemiotics with pH levels. Even more dramatically, the increasingly acidic seas are dissolving the shells of sea animals. Nina Bednaršek has documented the thin, partially dissolved shells of pteropods, tiny marine snails, which are "important as food for other zooplankton, fish, and marine mammals."[67] Many marine species, from krill to whales, depend on the pteropods, or sea butterflies, for food. If pteropods disappear from the polar and subpolar regions (to focus on just two regions), "their predators will be affected immediately": "For instance, gymnosomes are zooplankton that feed exclusively on shelled pteropods. Pteropods also contribute to the diet of diverse carnivorous zooplankton, myctophid and nototheniid fishes, North Pacific salmon, mackerel, herring, cod and baleen whales."[68] Pteropods are also important "biogeochemically," as part of the carbon cycle, when their shells sink to the ocean floor after their demise.[69] Considering how these creatures are crucial not only for the food web that sustains a multitude of other marine species but also as a carbon sink underscores the swirling, intimate interrelations between matters of biology, ecology, geology, and chemistry.

Whereas increasingly acidic seawater is itself difficult to represent in compelling ways, aesthetically entrancing images of dissolving shells of marine animals may enlist concern for ocean acidification. Nina Bednaršek's beautiful micrographs of two pteropod shells, one intact and one in the process of dissolving, appear in Ogden's article "Marine Life on Acid," but they also appear in the National Resource Defense Council's online magazine *On Earth*, the National Climate Assessment report, and the online technology publication *Ars Technica*.[70] Time-lapsed videos or photographs set in a series depict these dissolves in palpable manners. One striking panel of five images showing the pteropod shell dissolving at zero, fifteen, thirty, and forty-five days, by David Littschwager, which *National Geographic* owns the rights to as a stock image, appears on the NOAA website and on many other sites, including that of the Ukrainian Science Club.[71] Interestingly, in these images the actual fleshy creature that inhabits the shell is absent. The empty shells suggest that the animals did not survive, but they also may invite viewers to imagine taking up residence there, within the precarious abodes. The design of the shells, the spirals that swirl with a continual, smooth transformation between what is inside and what is outside, suggests the contemplation of our own bodies as intertwined with our surroundings.

Video depictions of dissolving shells are even more irresistible than the photographs. Julia Whitty includes Tim Senden's video of an X-ray micro-CT of a shelled pteropod *Limacina helicina antarctica* in her essay "Snails Are Dissolving in Acidic Ocean Waters," published in *Mother Jones* (Figure 10). The silent, twenty-two-second black-and-white video, which depicts a spinning, spiral, white shell, its edges dissolving into a transparent cloud, is rather entrancing, inviting a kind of mind-altering contemplation.[72] The beauty and fragility of the rotating shell are difficult to abandon. The brief black-and-white video is addictive. Highly mediated, depicting the shells of creatures rather than their fleshy bodies, these images nonetheless make claims on their viewers, seducing us to mobilize concern in scale-shifting modes.

The Tasmanian artist Melissa Smith, who creates art about the effects of climate change, has made several works featuring the pteropod within her Dissolve and Dissolve II series, including "Dispel," a stunning 2:30 video, animated by the same Tim Senden who produced the black-and-white video discussed above. The name, "Dispel," suggesting both dispersion and vanishing, shows a milky and translucent shell against a vibrant red background. Smith describes the video: "This work is emotively charged both visually and aurally. The cascading image of an X-ray micro-CT scanned pteropod shell, rotates and reveals its beauty before falling away to its demise. The soundtrack extends the viewer's perception of the visual to evoke an even deeper sense of loss."[73] The video begins with the shell gently falling into the frame of the camera and slowly, hypnotically rolling across the screen. Then it gets closer to the viewer, both encompassing the viewer, pulling her gaze in and through the spiral, but also allowing her to see through the transparency. The shell's extraordinary fragility is accompanied by mournful cello and piano music. In the end, revolving still, it disappears, white vanishing into red, as the shell spirals into smaller dimensions. The red background, signaling urgency, collides with the somber music and slow, mesmerizing rotations. The viewer's experience shifts from being a spectator, to being ensconced, to being part of the dissolve, left hovering within the red.

These shells, bereft of their fleshy creatures, without a face, nonetheless evoke concern, connection, empathy. While a gory scene depicting the living creature meeting its demise would separate the human spectator from this already distant form of marine life by sensationalizing it or rendering it abject, the elegant minimalist aesthetic of the shell lures us into a pleasurable encounter that nonetheless gestures toward the apocalyptic. Within the contemporary digital landscape in which ocean creatures are posed in highly aesthetic ways, by environmental organizations, scientists, and popular media, the shells take up their place in the virtual gallery of aesthetic marine pleasures, haunted by the missing fleshy life.[74] To say they call us to contemplate our own "shells"—or bodily and psychic boundaries—on acid, suggests something akin to a psychedelic experience. The spiral

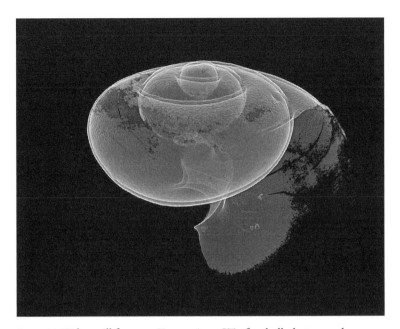

Figure 10. Video still from an X-ray micro-CT of a shelled pteropod
Limacina helicina antarctica. Courtesy of Tim Senden, Australian National
University CT Lab.

shells, especially when they are spinning around in the video versions,
do, in fact, suggest the spiral as the icon of altered states. This mode
of engagement, this type of attention, often involves a "dissolution
between" the human and the "outside world," as Wikipedia tells us:
"Some psychological effects may include an experience of radiant
colors, objects and surfaces appearing to ripple or 'breathe,' colored
patterns behind the closed eyelids (eidetic imagery), an altered sense
of time (time seems to be stretching, repeating itself, changing speed
or stopping), crawling geometric patterns overlaying walls and other
objects, morphing objects, a sense that one's thoughts are spiraling
into themselves, loss of a sense of identity or the ego (known as 'ego
death'), and other powerful psycho-physical reactions. Many users experience a dissolution between themselves and the 'outside world.'"[75]
Intrepid viewers may dis/identify in the dissolve, simultaneously identifying with the shelled creature and contemplating the dissolution of

boundaries that shore up human exceptionalism, imagining this particular creature's life and how extinction will ripple through the seas. This dissolution between the human self and the world suggests what Richard M. Doyle, in *Darwin's Pharmacy: Sex, Plants, and the Evolution of the Noösphere*, defines as an "ecodelic insight," "the sudden and absolute conviction that the psychonaut is involved in a densely interconnected ecosystem for which contemporary tactics of human identity are insufficient."[76] Although Doyle is not writing about the question of scale in terms of the anthropocene, his conception of the ecodelic may be useful for forging environmentally oriented conceptions of the anthropos, not as a bounded entity, nor as an abstract force, but as manifestation: "And in awe we forget ourselves, becoming aware of our context at much larger—and qualitatively distinct—scales of space and time. And over and over again we can read in ecodelic testimony that these encounters with immanence render the ego into a non sequitur, the self becoming tangibly a gift manifested by a much larger dissipative structure—the planet, the galaxy, the cosmos."[77] I am interested in how the ecodelic erodes the outlines of the individual self in "encounters with immanence" that provoke alluring modes of scale shifting. The problem here, however, is that contemplative or psychedelic practices have an association, in Western culture at least, with a navel-gazing, spiritual transcendence— the exact opposite of the sort of materially immersed subjectivity I think is necessary for environmentalism. Recasting Doyle's scenario by imagining the anthropogenically altered, acidified seas, rather than the perfect, ethereal expanses of the cosmos (descending, rather than transcending), may provoke a recognition of life as always immersed in substances and chemistries, that are, within the anthropocene especially, neither solid nor eternal. More difficult to contend with, however, is that the ecodelic figuration of the dissolve may be useless in terms of social justice and climate justice, in that it does not provoke consideration of differential human culpabilities and vulnerabilities. And yet, as a vivid image of slow violence, it could be taken up as a mode of dis/identification and alliance for particular groups of people who are contending with other sorts of invisible environmental harm. In her essay on the New Zealand Maori writer Kerri Hulme, Elizabeth DeLoughrey states that Hulme's stories "suggest that experience of embodied thought allows for merger with other species." DeLoughrey argues against apocalyptic fiction, however, and the figuration of the dissolve is rather apocalyptic. But other similarities resonate, such as her reading of Hulme's "narrative merger with fossils (and later the sea)" as "an encounter with deep planetary time that renders an interspecies relationship."[78]

As one figuration of the anthropocene among many others that are possible, the exquisite photographs and videos of dissolving shells may perform cultural work, portraying the shift in alkalinity as a vivid threat to delicate yet essential living creatures. Whereas the predominant sense of the anthropocene subject, en masse, is that of a safely abstracted force, the call to contemplate your shell on acid cultivates a fleshy posthumanist vulnerability that denies the possibility of any living creature existing in a state of separation from its environs. The image of the diminutive creature, with its delicate shell dissolving, provokes an intimacy, a desire to hold and protect, even as we recognize that such beings hover as part of the unfathomable seas. The scene of the dissolve demands an engaged, even fearsome activity of scale shifting from the tiny creature to the vast seas. In *The Posthuman*, Rosi Braidotti challenges us to imagine a vital notion of death: "The experiment of de-familiarization consists in trying to think to infinity, against the horror of the void, in the wilderness of non-human mental landscapes, with the shadow of death dangling in front of our eyes."[79] Arguing not for transcendence but instead for "radical empirical immanence," she contends that "what we humans truly yearn for is to disappear by merging into this generative flow of becoming, the precondition for which is the loss, disappearance and disruption of the atomized, individual self."[80] Envisioning the dissolve, then, can be an immanent, inhuman or posthuman practice.

In the era of the sixth great extinction, it is not difficult to discern the shadow of death. Marine life faces many other threats in addition to acidification, including warming waters and the ravages of mining, drilling, ghost nets, shark finning, and industrial overfishing. Marine habitats are riddled with radioactive waste, toxic chemicals, plastics, and microplastics, all of which become part of the sea creatures that, not unlike Beck's citizen in risk society, lack the means to discern danger, and the impermeability that would exclude it. Contemplating your shell on acid is a mode of posthumanist trans-corporeality that insists all creatures of the anthropocene dwell at the crossroads of body and place, where nothing is natural or safe or contained. To ignore the invisible threats of acidity or toxins or radioactivity is to imagine we are less permeable than we are and to take refuge in an epistemological and ontological zone that is somehow outside the time and space of the anthropocene. Those humans most responsible for carbon emissions, extraction, and pollution must contemplate our shells on acid. This is a call for scale shifting that is intrepidly—even psychedelically—empathetic, rather than safely ensconced. Contemplating your shell on acid dissolves individualist, consumerist subjectivity in which the world consists primarily of externalized entities, objects for human consumption. It means dwelling in the dissolve, a dangerous pleasure, a paradoxical ecodelic expansion and dissolution of the human, an aesthetic incitement to extend and connect with vulnerable creaturely life and with the inhuman, unfathomable expanses of the seas. It is to expose oneself as a political act, to shift toward a particularly feminist mode of ethical and political engagement.

CONCLUSION
Thinking as the Stuff of the World

◇◇

Just a few lines from Jorie Graham's poem "Sea Change" evoke anxiety about unpredictable futures that arrive too soon, in need of repair:

> mornings in the unknown future. Who shall repair this. And how
>   the future
> takes shape
> too quickly. The permanent is ebbing. Is leaving[1]

The abrupt departure of a sense of permanence may provoke the desire to arrest change, to shore up solidity, to make things, systems, standards of living "sustainable." The call in the prior chapter to contemplate one's shell on acid and dwell in the dissolve needs to be contextualized as a radical departure from the most influential version of "environmentalism" of the last several decades—that of sustainability. Having worked in the environmental humanities and science studies for more than a decade and having served as the academic cochair for the University Sustainability Committee at the University of Texas of Arlington for several years, I have been struck by how the discourse of "sustainability" at the turn of the twenty-first century in the United States echoes that of "conservation" at the turn of the twentieth century, especially in its tendency to render the lively world as a storehouse of supplies for the elite. Gifford Pinchot, Theodore Roosevelt's head of forestry, defined forests as "manufacturing plants for wood," epitomizing the utilitarianism of the conservation movement of the Progressive era, which saw nature as a resource for human use. By the early twentieth century Pinchot's deadening conception of nature jostled with other ideas, such as those of aesthetic conservation and the fledgling science of ecology. Pinchot was joined by the Progressive Women Conservationists who claimed, as part of the broader "Municipal Housekeeping" movement, that women had special domestic talents for conservation, such as "turning yesterday's roast into tomorrow's hash."[2] Many Progressive Women Conservationists not only bolstered traditional gender roles, but also wove classism and racism into their conservation mission, as "conservation" became bound up with conserving their own privileges. The anthropocentrism of the Progressive Women Conservationists is notable. As one participant at the First National Conservation Congress stated in 1909, "Why do we care about forests and streams? Because of the children who are to be naked and bare and poor without them in the years to come unless you men of this great conservation work do well your work." During their conventions, the discourse of conservation was playfully and not so playfully extended to myriad causes, including conserving food, conserving the home, conserving morals, conserving "true womanliness," conserving "the race," conserving "the farmer's wife," and conserving time by omitting a speech.[3]

The U.S. frenzy to conserve at the turn of the twentieth century was, in part, driven by the desire to demarcate the country's resources as belonging to some groups and not others as waves of immigrants came ashore. The current mushrooming of the term "sustainability," too, may be fueled by anti-immigration fervor, as well as by the desire to entrench systemic inequalities during a time of economic instability. At the start of the twenty-first century, anti-immigration movements focusing on the southwestern border of the United States are complemented by anxious glances toward the East, as the economies of China, East Asia, and India expand. Fear lurks behind the proliferating, sanitized term "sustainability," as news reports worry that economies, national debts, personal debt, the housing market, food systems, the Eurozone, and all manner of more trivial matters are not "sustainable." Although the concept of sustainability emerges, in part, from economic theories that roundly critique the assumption that economic prosperity must be fueled by continual growth, the term is frequently invoked within economic and other news stories that do not, in any way, question capitalist ideals of unfettered expansion.[4] Like "conservation," sustainability has become a plastic but potent signifier, meaning, roughly, the ability to somehow keep things going, despite, or rather because of, the fact that we suspect economic and environmental crises render this impossible. In other words, "sustainability" reveals the desire for inertia, propelled by denial. John P. O'Grady points out the irony here: "That nothing stays the same is the very basis of history [and] evolutionary theory." Thus "there is no ecological justification for the idea of sustainability."[5] The discursive success of the signifier—in business, science, academics, and popular culture—leads one to suspect that it may be serving a psychological function in the social consciousness. Although Slavoj Žižek in *Living in the End Times* does not dwell on sustainability, he does analyze the mechanisms that allow us to maintain ourselves psychologically while an apocalypse gallops toward us. For example, we "*know* the (ecological) catastrophe is possible, probable even, yet we do not *believe* it will really happen."[6] Could dwelling in the dissolve suspend this disbelief?

Disciplining Movements, Academics, and Knowledges

Even as the movement for more sustainable universities, businesses, cities, states, and households is a positive development, in that the systematic attempt to reduce energy and water usage, reduce waste, use less toxic products, and shrink carbon footprints is nothing to dismiss, we may well ask how it is that environmentalism as a social movement became so smoothly institutionalized as "sustainability." The discourse of sustainability, cleansed of its association with "tree huggers," and articulated to a more technocratic, apolitical domain, is more palatable for academic institutions, governments, and businesses. While it would be politically awkward for colleges and universities to ally themselves with environmentalism per se, which may offend some donors or legislators, 840 institutions of higher education are members of the Association for the Advancement of Sustainability in Higher Education. On university campuses such things as Environmental Management Systems, defined by the U.S. EPA as "a set of processes and practices that enable an organization to reduce its environmental impacts and increase its operating efficiency,"[7] complement the growing faculty management systems in which academic labor must not only become more "efficient" but must be measured in ever more quantitative ways. Not surprisingly, this new gospel of efficiency[8] values the disciplines that can *fix* things—engineering, the sciences, and maybe architecture and urban planning. Who has time for philosophical questions, social and political analyses, historical reflections, or literary musings when the world is rapidly heating up and "resources" are running out?

The humanities may be dismissed outright when it comes to the "triple bottom line" of profit, people, and planet. Stephanie LeMenager and Stephanie Foote advocate for what they term the "sustainable humanities," which denotes, broadly, the "ecological value of humanities education," by cultivating intergenerational memory and a deprivatized civic sphere.[9] Daniel J. Philippon, in "Sustainable Humanities: An Extensive Pleasure," lists eight tools that the humanities offer sustainability—defining, theorizing, imagining, specifying, categorizing, historicizing, contextualizing, comparing—and then, he doubles this list by insisting each of these modes be accompanied by questioning—questioning definitions, theories, categories, and so forth.[10] The fact that the Institute for Humanities Research at Arizona State University drafted a white paper titled "Contributions of the Humanities to Issues of Sustainability" suggests that these contributions require explanation. The first of the seventeen contributions in this convincing document asserts that the humanities are crucial for both understanding and solving environmental crises, as humanists "challenge reliance upon the authority of 'nature' or 'science' in order to address problems that in their origin and solution are primarily social and cultural."[11] Gert Goeminne would agree with this assertion. In "Once upon a Time I Was a Nuclear Physicist: What the Politics of Sustainability Can Learn from the Nuclear Laboratory," Goeminne argues that "expert-focused technological determinism, embedded in a discourse of ecological modernization, now acts to marginalize the issues of human choice involved in putting sustainability into effect and to downplay deliberation over the socio-cultural practices, behaviours, and structures such choice involves. As a result of this techno-scientific focus, the need for accordant social change is removed from view, which makes sustainability all the less likely to occur in practice."[12] This technological focus obscures power differentials, political differences, cultural values, and everyday human practices. Technical problems and their solutions become compressed and contained. This may yield some valuable inventions and designs, but the wider human, geographic, social, political, and economic contexts and interconnections are obscured, along with an understanding of multiple scales. The narrow, technical notion of sustainability could be countered by the sort of critical realism that LeMenager and Foote contend humanities scholarship and pedagogy are well positioned to provide: the artful representation of "realities that are not usually visible because of the scalar extremes and privatization of space in capitalism today."[13]

The techno-scientific perspective contrasts with the alternative epistemologies of "popular epidemiologists" and "ordinary experts" that have emerged from environmental justice and environmental health movements.[14] As citizens with little or no scientific background take samples and analyze data gleaned from their own communities, science is shown to be a politically embedded practice that is too important to be left to the experts. Environmental justice groups are not alone, however, in challenging traditional models of scientific distancing, objectivity, and authority. Environmental health movements, people with multiple chemical sensitivity, domestic carbon footprint analysts, and environmental activists of all sorts practice DIY (do-it-yourself) science. The chemically sensitive move through the world using their own bodies as monitoring devices, treesitters in the Pacific Northwest from their vantage point hundreds of feet in the air assess how clear-cutting leads to mudslides, and dolphin advocates on the Texas coast monitor the behavior, communication, and kinship patterns of cetaceans.[15] These modes of knowledge are embedded, passionate, and purposeful—the mirror image of scientific objectivity. Even as the practices of sustainability foster the recognition that nearly everything one does has effects on larger environmental issues, the epistemological stance of sustainability, as it is linked to systems management and technological fixes, presents a rather comforting, conventional sense that the problem is out there, distinct from one's self. The dominant style of sustainability parallels that of the anthropocene vision discussed in the prior chapter, in which a disembodied spectator is outside the externalized and inert world. Sustainability proceeds with the presumption that human agency, technology, and master plans will get things under control. But the crises of the anthropocene render that stance absurd, as the unintentional effects of human activity, and its interactions with other forces, outpace even the best laid plans. Throughout *Exposed*, embedded onto-epistemologies, provisional knowledge practices, performances of exposure, and imaginative dissolves diverge from the predominant paradigm of sustainability by staying low, remaining open to the world, and becoming attuned to strange agencies.

Rosi Braidotti embraces the possibilities for the concept of sustainability, arguing that what it stands for is "a regrounding of the subject

in a materially embedded sense of responsibility and ethical accountability for the environments she or he inhabits."[16] Braidotti infuses
sustainability with a Deleuzian sense of becoming: "The ethical subject of sustainable becoming practices a humble kind of hope, rooted
in the ordinary micro-practices of everyday life: simple strategies to
hold, sustain and map out thresholds of sustainable transformation."[17]
This affirmative sense of transformative micro-practices is countered
by Žižek's condemnation of such activities as purchasing organic food
as yet another mode of disavowal: "I know very well that I cannot
really influence the process which may lead to my ruin (like a vol-
canic eruption), but it is nonetheless too traumatic for me to accept
this, so I cannot resist the urge to do something, even if I know it
is ultimately meaningless."[18] Between Braidotti's humble yet utopian
sense of transformation and Žižek's impotent activities of disavowal
dwell the less exuberant and less certain practices of environmental
justice activists and amateur practitioners, who recognize that their
own bodily existence is caught up in material agencies that are difficult
to discern, and often impossible to escape.

While the epistemological stance of sustainability offers a comforting sense of scientific distancing and objectivity, trans-corporeal
subjects are often forced to recognize that their own material selves
are the very stuff of the agential world they seek to understand. The
literary and popular genre of what I term the "material memoir,"
most notably Susanne Antonetta's *Body Toxic*, for example, transforms
autobiography into an examination—often scientific—of how the self
is coextensive with the environment.[19] Similarly, while the promotion
of, say, "sustainable seafood" holds out the possibilities that there are
marine creatures that can be consumed without threatening their continued existence or harming the health of those who eat them, the
activist short film "A Shared Fate," discussed in chapter 5, documents
how mercury and PCBs not only kill massive numbers of dolphins and
other marine mammals but also threaten humans who eat dolphins
and whales, as well as those humans who consume the same fishes
that dolphins and whales consume. The video reveals that Hardy
Jones, who had devoted his life to protecting cetaceans from slaughter,
ends up suffering from the same form cancer that is killing them, as
his own body carries high levels of mercury and other heavy metals.
An appeal to "sustainability" would be a rather abstract and ineffectual
gesture for this drama that demands, at the very least, more stringent

movements and measures to prevent massive amounts of mercury and toxic chemicals from entering the oceans. And while tracing trans-corporeal toxic flows and material embeddedness may sound like rather dismal onto-epistemological practices, Philippon, drawing on Kate Soper's "alternative hedonisms," and Wendell Berry's notion of ethical modes of eating as an "extensive pleasure," urges us to counter the joylessness of environmentalism with practices that are simultaneously intellectual, political, aesthetic, and pleasurable. Philippon's sense of sustainability as potentially pleasurable parallels some of the arguments for pleasure throughout this book. He also offers a striking account of intellectual engagement, arguing not only that "the pursuit of academic pleasure cannot be separated from the context of that pursuit," but that "the meaning of that pleasure grows when its political implications are made clear."[20] Extensive pleasures then, as academic and disciplinary practices, can help sustain the humanities as the humanities contribute to larger projects of sustainability.

Material Agencies and Posthuman Futures

Scholars in the humanities, or, more aptly, the posthumanities, may well ask, "What is it that sustainability seeks to sustain?" and "For whom?" Questions of social justice, global capitalist rapacity, and unequal relations between the global north and the global south are invaluable for developing models of sustainability that do more than try to maintain the current, brutally unjust status quo. Julian Agyeman, stressing the "inseparability of environmental quality and human equality," promotes the concept of "just sustainability," which focuses on race and class and involves redistribution and transformation.[21] While I would agree that sustainability movements must integrate social justice, and that income disparities call out for redistribution, the focus on redistribution within a sustainability model still poses nature as inert resources for human use and ignores multispecies claims. Moreover, many environmental harms are silent, invisible, and difficult to detect. Concepts of sustainability, even just sustainability, need to be troubled by the recognition of the pervasive "slow violence," in Rob Nixon's terms, that characterizes the "environmentalism of the poor."[22] Imagine that environmental management system metrics included data on "violence," "disease," "genetic damage," or "death," gleaned from the long-term impact of "resource" extraction, manufacturing, use, and disposal. The "resources" act in unruly ways. The absence of this sort of data is accompanied by other glaring absences. Strangely, "sustainability" evokes an environmentalism without an environment, an ecology devoid of other living creatures. The standard definition of sustainability, given in the 1987 Brundtland Report is as follows: "Development that meets the needs of the present without compromising the ability of future generations to meet their own needs."[23] Not only are the "generations" here usually taken to be human but the lively world is reduced to the material for meeting the "needs" of future humans. ("Why do we care about forests and streams? Because of the children . . .")

The anthropocentric rhetorics of sustainability echo those of climate change movements. As chapter 4 discussed, climate change initiatives by the United Nations and global feminist NGOs are strangely devoid of nonhuman creatures. Even Bill McKibben's 350.org, "dedicated to building a global grassroots movement to solve the climate crisis," features photos of assemblies of people around the world who hold up signs or flags or wear T-shirts displaying the number "350," the "'safe upper limit' of carbon dioxide in our atmosphere" in parts per million. The photographs depict large groups of people—flying kites, biking, walking, marching, standing, or arranged in symbolic shapes. Absent are the less cheery images one would expect from a climate change movement, such as photographs of agricultural areas ravaged by drought, clear-cut forests, oil- and sludge-filled Amazonian regions, and melting icebergs. Instead, we see happy, smiling faces gathered on behalf of a number, 350. As picture after picture continues to roll by, there is not a nonhuman animal—wild or domesticated—in sight. When McKibben spoke at the University of Texas at Arlington in 2010 he showed these same sorts of photographs, a veritable parade of equalized global peoples. He did not mention how climate change is expected to condemn a million species to extinction by 2050. He did not show photographs of any of these species or their habitats. The nonhuman species seem to have already disappeared, at least in terms of visibility or concern. The animated feature on the 350.org site, designed, laudably, to explain the movement without using language, so as to be understood by any human anywhere, depicts human stick figures moving on blank, lifeless backgrounds. Humans and their activities are animated, but the material world is rendered as abstract space, not living places, biodiverse habitats, or ravaged ecologies. An invisible *any where*. McKibben, famous for his eponymous declaration in *The End of Nature* that nothing can now be called nature (if "nature" suggests something untouched by the human),[24] has, disturbingly, set his climate change movement in a world utterly devoid of other than human life-forms, agencies, habitats, and systems.

In his afterword to *Living in the End Times*, Žižek calls us to consider the commons: "Communism is today not the name of a solution but the name of a problem: the problem of the commons in all its dimensions—the commons of nature as the substance of our life, the problem of our biogenetic commons, the problem of our cultural commons ('intellectual property'), and, last but not least, the problem of the commons as that universal space of humanity from which no one should be excluded."[25] Is it possible to imagine this "universal space of humanity" as including all nonhuman life-forms, in a manner akin to Bruno Latour's collective of humans and nonhumans? Doubtful. Nonhumans, it seems, have already been excluded from the space "from which no one should be excluded." The lively, agential world of diverse creatures becomes a blank, "universal space," or the "substance" of "our life." The tragedy of Žižek's commons is that they are exclusively, sadly, *ours* alone. The challenge of how to include the claims, needs, and agencies of other living creatures, habitats, and ecosystems remains.

Has the term "sustainability" become articulated too firmly to a technocratic, anthropocentric perspective? Is it possible to recast "sustainability" in such a way that it ceases to epitomize distancing epistemologies that render the world as resource for human use? Should biodiversity be one of the principal, or even the foremost, states to be "sustained," notwithstanding the fact that perpetual change, not fixity, is the ungrounded ground for the survival of diverse species? Could sustainability be transformed in such a way as to cultivate posthumanist epistemologies, ethics, politics, and even aesthetics? Consider the oceans. The very liquidity of pelagic habitats, alien to human understanding, may dislodge us from our entrenched way of approaching the world. Denise L. Breitburg and colleagues in "Ecosystem Engineers in the Pelagic Realm: Alteration of Habitat by Species Ranging from Microbes to Jellyfish," for example, explain that many species in the open waters of the seas "completely transform the pelagic habitat."[26] The cumulative effects of "many small actions" result in "habitat that is created or altered over large spatial and temporal scales."[27] Considering creatures from microbes to jellyfish as themselves "ecosystem engineers" stresses the lively interactions within watery worlds while underscoring that the physical environment is never mere "background" or abstract space.

Recognizing that the "permanent is ebbing" and the "unknown future" will surely be in need of "repair" may discourage us from taking refuge in the idea that we can fix the world out there in such a way as to ensure "it" will keep providing for "us." Perhaps what the environmental humanities and science studies can contribute to "sustainability," if indeed we choose to use that term, would be to formulate more complex epistemological, ontological, ethical, and political perspectives in which the human can no longer retreat to an asylum of separation and denial, or to proceed as if it were possible to secure an inert, discrete, externalized this or that. New materialist conceptions of agency, explicitly put forth in theory and manifest in various modes of environmentalist activism, strike to the root of what is wrong with the concept of sustainability. Barad's posthuman ethics, for example, counters the tendency of sustainability to externalize and objectify the world through management systems and technological fixes. Barad argues that an "ethics of mattering" "is not about right response to a radically exterior/ized other, but about responsibility and accountability for the lively relationalities of becoming of which we are a part."[28] The interacting material agencies provoked by the staggering scale and fearsome pace of human activities will no doubt bring about unknown futures. Rather than approaching this world as a warehouse of inert things we wish to pile up for later use, we must hold ourselves accountable to a materiality that is never merely an external, blank, or inert space but the active, emergent substance of ourselves and others.

Thinking as the Stuff of the World

Being invited to write for the inaugural issue of *O-Zone: A Journal of Object-Oriented Studies* forced me to consider the parallels and the distinctions between different movements within the material turn in theory. As a new materialist and material feminist who developed the concept of "trans-corporeality," which foregrounds material agencies, I understand the necessity for theory and cultural criticism to forge new ways of accounting for the agencies and significance of material substances, forces, and systems. But as an ecocultural and animal studies theorist I bristle at the first word of object-oriented ontology (OOO)—"object"—which erases all distinctions between consumer projects and living creatures. Yet haven't the last few decades of science studies, feminist theory, and other fields reconfigured the established divides between subject and object, nature and culture, in such a way to result in the flat ontology of "objects"? Are the objects of OOO fundamentally different from Haraway's oncomouse or Latour's hole in the ozone? Levi Bryant explains that a flat ontology seeks to overcome human exceptionalism: "Rather than bifurcating being into two domains—the domain of objects and the domain of subjects, the domain of nature and the domain of culture—we must instead conceive of being as a single flat plane, a single nature, on which humans are beings among other beings."[29] It has been difficult for theorists to formulate language that can cross the nature/culture divide, as most Western terminology is always already rooted to one side or the other. The term "equality," for example, adheres to the political and social realm.[30] When Ian Bogost contends that "nothing has special status, but that everything exists equally—plumbers, cotton, bonobos, DVD players, and sandstone, for example,"[31] I wonder, why place bonobos and DVD players and plumbers on an equal plane? Doesn't this flat plane quash the animal studies arguments for animal minds, animal cultures, animal communications? (Sure, there is a plumber on that list, but there is little danger that standing adjacent to cotton will dismantle sturdy humanist presumptions.) Is the focus on objects too posthumanist or not posthumanist enough? How would (how do) the philosophical interventions of OOO play out in popular culture, politics, activism, and daily life? What is the relation between the objects of OOO and consumer products? And what are the relations between emerging transdisciplinary fields and movements such as the nonhuman turn, thing theory, new vitalism, speculative realism, affect theory, new materialism, material feminisms, posthumanism, animal studies, and OOO? This is not the place to address all these questions of course, so I will focus here on how OOO contrasts with the positions, predilections, politics, and arguments of this book.

In *Alien Phenomenology; or, What It's Like to Be a Thing* Bogost dismisses environmental studies, animal studies, science studies, and posthumanism, within a mere two pages, for not being posthumanist enough. But he cites very few examples, claiming, mistakenly, that "posthuman approaches still preserve humanity as primary actor" and that science studies "[retain] some human agent at the center of analysis."[32] He also charges, correctly, but oddly, that environmentalism limits its concern to "living creatures."[33] By ignoring the work of feminist science studies scholars such as Donna Haraway, Nancy Tuana, and Karen Barad, he positions OOO as the only escape route from the "tiny prison of our own devising," in which "all existence is drawn through the sieve of humanity."[34] Barad's theory of agential realism, for example, offers a potent conception of material agency that does not privilege the human. Moreover, Barad challenges the very notion of discrete "objects." Bogost's lists of things—which mix types of humans with animals, household appliances, banal consumer products, and anything else one could think of—circumscribe each *thing* as a separate entity. Barad, on the other hand, drawing on Neils Bohr, takes "the primary ontological unity to be phenomena, rather than independent objects with inherent boundaries and properties. . . . *Phenomena are the ontological inseparability of intra-acting agencies.*" That is, "*phenomena are ontological entanglements.*"[35] Inhabiting Barad's theory, contemplating the utterly counterintuitive sense of the world as made up of intra-acting agencies, rather than separate objects, is to me more vertiginous and generative than contemplating objects as distinct alien beings. Moreover, the focus on detached objects promotes a consumerist ideology, especially as capitalism would like nothing better than to utterly obscure relations of production, in which the object is entangled with exploitative working conditions and environmental harms. By contrast, tracing intra-actions across substances, systems, and bodies enables political critique, economic interventions, and the development of less harmful practices.

Levi Bryant, taking up Bogost's conception of "alien phenomenology," contends that "in all cases" it "consists in the attempt to suspend our own human ways of operating and encountering the world so as [to] investigate non-human ways of encountering the world."[36] This is a valuable endeavor for posthumanism, animal studies, and plant studies, and Bryant's conception of the "machine" rather than the object levels out the human and nonhuman in a way that eludes anthropocentrism while also escaping "a highly sedimented philosophical tradition surrounding objects and subjects."[37] Bryant also insists on the ethical and political potential of alien phenomenology, contending that "a great deal of human cruelty arises from the failure to practice alien phenomenology," including sexism, "colonial exploitation, op-

pression, and genocide," and the mistreatment of animals. He proposes that "through the practice of alien phenomenology, we might
develop ways of living that are both more compassionate for our others and that might develop more satisfying social assemblages for all
machines involved."[38] Bogost, however, disregards the environmental
and social implications of these theories, as he reinstalls a humanist
and masculinist disembodied subject. The philosopher asks, "What is
it like to be a computer or a microprocessor, or a ribbon cable? . . . As
operators or engineers we may be able to describe how they *work*.
But what do they *experience*? What's their proper phenomenology? In
short, what is it like to be a thing?"[39] He rejects science studies as the
route to an answer because it "retains some human agent at the center of analysis."[40] But the method of philosophical speculation seems
terribly ill equipped for the task of accessing objects, as it places the
human mind squarely in the "center of the analysis." Allow me to
note that I cannot drink the Kool-Aid here and believe that a cable
experiences anything at all; nor do I find it useful—personally, intellectually, ethically, politically, or in any other way except for perhaps
as some sort of psychedelic koan—to imagine what it is like to "be"
a cable. I do wonder, however, albeit rather anthropocentrically, what
it is like to be a human imagining what it means to be a thing. In
this case, Bogost's speculations on what it means to be a particular
object emerge from a detached, rational mind.[41] There is no sense of
embodied, interactive, intra-active, situated, or scientifically mediated
knowledges here. Feminist, postcolonial, and environmental epistemologies have long critiqued modes of knowing that install a gap between the subject and the object of knowledge. But these theories are
overlooked. Instead, the knower who undertakes the phenomenological explorations of the aliens that surround him is separate from that
which he ponders. Since I do not expect a cable to possess a sense of
"being," it is not surprising that we are not given a vivid account of
what it is to be a cable. But it is strange that this alien phenomenology
ends up telling us so little about the cable or any other object. The
abyss between the philosopher and the objects he contemplates, leads
us to Derrida. If we follow Derrida following the animal, we might
trace some parallels here, as the philosophers' point of enunciation is
that of being those who "have given themselves the right to give" the
word—in this case, the word would be that of "object," rather than
"animal": "the word that enables them to speak of the animal with a

single voice and to designate it as the single being that remains without a response, without a word with which to respond."[42] Although OOO intends to level various entities, putting the human on the same ontological plane as other "objects," the human voice is the only thing we hear.

Philosophical contemplation may not be the most generative method for accessing objects, substances, or other materialities. Andrew Pickering's model of the "mangle of practice" is a revealing counterpoint here, as it allows for nonhuman agencies to register, even as it accounts for the interactions between the scientific, economic, and social.[43] Kim TallBear's extraordinarily promising, multidimensional project on the scientific, regulatory, and indigenous approaches to pipestone provides another striking model, which approaches the material as that which is vitally interconnected with lives, stories, and practices. She counters the way "making monuments and doing science risk deanimating" pipestone, by analyzing the stone from Dakota standpoints, where it is more like a relative than an alien:

> The stone is sometimes spoken of as a relative. Unlike with blood or DNA, pipestone does not possess a cellular vibrancy. Yet without it, prayers would be grounded, human social relations impaired, and everyday lives of quarries and carvers depleted of the meaning they derive from working with stone. Just like indigenous people who insist on their continuing survival and involvement with their DNA, indigenous quarries and carvers, medicine people, and everyday people who pray insist on living with the red stone daily.[44]

TallBear's methods, which involve science, site-based research, archives, and participant observation, promise to yield rich, robust analyses as she develops "indigenous, feminist, and queer theory approaches to critical 'animal studies' and new materialisms."[45]

Readers might object that the prior chapter in this book does something similar to cable contemplation. Dwelling in the dissolve of the acidifying seas also involves an imaginative, speculation across an abyss. But my call to consider your shell on acid entails a sense of the human self as permeable, part of the flux and flow of the anthropocene, part of the stuff of the world. It is a call to contemplate the actual acidification of the seas as scientific data are captured, to reckon with the moment of the sixth great extinction, and to inhabit an environmentally ethical sense of the self as immersed within an altering world. Trans-corporeality, as it reckons with material agencies that traverse substances, objects, bodies, and environments, entails reckoning with scientific captures, even as the data are always already mangled by social and economic forces.

I agree, as many environmentalists would, with Timothy Morton's contention that what "ecological thought must do, then, is to unground the human by forcing it back onto the ground," but disagree with what follows, "which is to say, standing on a gigantic object called *Earth* inside a gigantic entity called *biosphere*."[46] Defamiliarization affords aesthetic pleasure, of course, but the scalar leap from the ground inhabited by the human to the earth as "gigantic object" obscures the sort of entanglements that are the very stuff of ethical and political relations. Morton distinguishes his position from "ecophenomenology, which insists on regressing to fantasies of embeddedness."[47] But embeddedness need not be phenomenological, nor a regressive fantasy. Rather, the embeddedness of trans-corporeality involves grappling with data, information, scientific captures, and political modes of mapping interactions and relations across different scales. This requires too much work to be a regressive fantasy. The more convenient and persistent fantasy of the human, as we have seen throughout this book, is that he is free floating, unencumbered, and anything but embedded. Morton's ungrounded human who contemplates the earth as a "gigantic entity" seems similar to the human who views the predominant visual depictions of the anthropocene discussed in the prior chapter, in that they are both disembodied and disconnected from the scene. A rather different creature appears in *Hyperobjects* a few pages later, however, in a chapter called "Viscosity." Here Morton writes from his own embodiment. As global warming burns the skin on his neck, he tells of the agency of hyperobjects and the mercury and toxins in his blood.[48] This embodied human is more like that of the trans-corporeal subject, or the subject conceptualized by Nancy Tuana in "Viscous Porosity: Witnessing Katrina." Tuana argues that "in witnessing Katrina, the urgency of embracing an ontology that rematerializes the social and takes seriously the agency of the natural is rendered apparent."[49] She explains how drinking out of a plastic water bottle transforms her flesh: "Once the molecular interaction occurs, there is no divide between nature/culture, natural/artificial."[50] Plastics even interact with something as social as poverty: "Political failures to address the environmental hazards of plastics have left their signature on the flesh of many bodies, but the bodies of industry workers who toil in the plastics factories or the garbage incinerators and the bodies of those who live in the path of their pollutants have disproportionately suffered the negative effects of this material–semiotic interaction."[51]

Exposure, then, is terribly uneven, across such simultaneously social and material categories as class, race, and the disparities between the global north and the global south. And while much of this book has emphasized the material dimensions of the exposure, it is crucial to point out that ideological and discursive categories position bodies differently and have material effects. For feminists, LGBTQ people, people of color, persons with disabilities, and others, thinking through how corporeal processes, desires, orientations, and harms are in accordance with or divergent from social categories, norms, and discourses is a necessary epistemological and political process. For some people this is a matter of survival. Eli Clare in "Meditations on Natural Worlds, Disabled Bodies, and a Politics of Cure" writes first of the terms *"natural* and *unnatural, normal* and *abnormal"* as they are invoked within prairie restoration projects, and then states, "It is not an exaggeration to say that the words *unnatural* and *abnormal* haunt me as a disabled person. Or more accurately, they pummel me."[52] Clare works through the painful clashes between the discourses of disability[53] and environmental restoration, to arrive at the choice "between monocultures, on one hand, and bio- and cultural diversities on the other, between eradication and uncontainable flourishing."[54] Clare concludes with a clearly stated position, but what drives the essay is his more complex location within the material–semiotic landscapes of the language and practices of ecological restoration as well as the experience of living as a body that cannot be "restored."

Ideals of restoration, which hark back to stable ecosystems and natural genders, are unthinkable within Beatriz (Paul) Preciado's work. In *Testo Junkie: Sex, Drugs, and Biopolitics in the Pharmacopornographic Era* Preciado engages in DIY transformations of gender, sex, body, and mind by taking testosterone as a political act. Preciado's "theory" emerges from this praxis, from altering and experiencing the very "pharmacopornographic body" theorized, which is "not passive living matter but a techno-organic interface, a technoliving system segmented and territorialized by different (textual, data-processing, biochemical) political technologies." Preciado shifts from the idea that "the body inhabits disciplinary spaces" to a critical/theoretical/corporeal/political practice in which "the body is inhabited by . . . biopolitical systems of control."[55] This is a radical sense of trans-corporeality—one that underscores hormones, gender, sex, and pharmacology—potently demonstrating that the (post)human is never natural, never detached, never discrete. To analyze, theorize, critique, create, revolt, and transform as someone whose corporeality cannot be distinct from biopolitical systems and biochemical processes is to think as the stuff of the world.

Eva Hayward's dazzling work demonstrates how the long history of feminist and queer writing as a politics of exposure may flourish as a newly transfigured posthumanism. Her work, in my view, rises to the formidable challenge Cary Wolfe poses in *What Is Posthumanism?* when he states that the "nature of thought itself must change if it is to be posthumanist."[56] In "More Lessons from a Starfish: Prefixial Flesh and Transpeciated Selves," Hayward writes of transsexuality as a "mutuality," a "shared ontology," with the starfish:

> Trans-morphic as zoomorphic—if we can understand
> the cut as an act of love, then can we not imagine that
> "like a starfish" is an enactment of trans-speciating? We,
> transsexuals and starfish, are animate bodies; our bodies
> are experienced and come to be known through encoun-
> ters with other animate bodies. These epistemological
> moves describe a shared phenomenological ontology.
> This is sensate intertwining—intercorporeal zones
> between these bodies in language and in experience.
> Starfish and transsexuals share worldhood both semi-
> otic (as metonymic kinds) and phenomenological enact-
> ments—is this not some form of *inter*corporeality?[57]

Hayward, by thinking her body across the regenerations of the starfish, "fundamentally unsettle[s] and reconfigure[s] the question of the knowing subject," as Cary Wolfe puts it,[58] since the starfish—and the human self—are known through a kind of intercorporeality. This is a palpable, dazzling, posthumanist figuration, as the shared ontology with the starfish culminates in a "transspeciated self," a self who is, who knows, through an encounter with another species. Hayward describes her method as a "critical enmeshment rather than a personal account," arguing that the "material, the literal matter, of being, surfaces and resurfaces as a constitutive force that cannot be digested in the acid fluids of anthropic concerns."[59] Hayward's work exemplifies the possibilities for new materialist thought to emerge from lived genders, sexualities, and other embodied knowledges and performances of exposure. But it also exemplifies how thinking with a multitude of living creatures may enrich new materialist theory, and thus how fruitful the alliances may be between new materialisms, posthumanisms, and animal studies. It is also important to point out, however, that as speculative and creative and intrepid as Hayward's work is, she draws on scientific disclosures about starfish rather than simply imagining their being. Hekman argues that the concept of disclosure "avoids the problem of representationalism" and relativism, offering us a model, such as that of Andrew Pickering's mangle, in which "multiple elements interact, or intra-act, to produce an understanding of the reality we share."[60] Although that reality is shared, animal studies scholarship insists that particular species experience and understand the world in significantly different ways. Hayward, for example, explores how, although "many echinoderms do not have many well-defined sensory inputs, they are sensitive to touch, light, temperature, orientation, and the status of water around them": "their very *being* is a visual–haptic–sensory apparatus."[61]

One chapter in *Alien Phenomenology* is titled "Revealing the Rich Variety of Being," but there are no creatures akin to Hayward's starfish dwelling there, nor is there any mention of the sixth great extinction, which is, no doubt, diminishing the rich variety of being. The celebration of consumer objects as fascinatingly alien diverts attention from the loss of living creatures in the world. When Timothy Morton cavalierly states, "We might add that OOO radically displaces the human by insisting that my being is not everything it's cracked up to be—or rather that the being of a paper cup is as profound as mine,"[62] we might wonder what "profound" and "being" mean in this context. Such an argument may circulate as a sort of reductio ad absurdum of arguments for the rights, welfare, or continued existence of nonhuman creatures, as they leap over serious claims about how characteristics formerly reserved for the human, such as consciousness, intelligence, communication, and the capacity to suffer are in fact distributed throughout nonhuman animals and cultures—landing in an absurd reality where cups can be considered profound beings. A different sort of attention to consumer objects, percolating through environmental activism and new materialism, traces and reckons with the often invisible systems and networks that produce ordinary objects, discloses the often unintended consequences of the material agencies of those objects, and forges new ethical and political practices that arise from the material–semiotic entanglements of the world. It is how objects are entangled—economically, politically, and substantially across bodies, ecosystems, and built environments—that matters, not how each object exists in isolation.

I share Bogost's contention that "wonder has been all but eviscerated in modern thought." And the mundane–sublime of alien phenomenology, in which we celebrate "the awesome plentitude of the everyday,"[63] offers a Zen-like appreciation for what is, giving us respite from anguished worries about what (or who) will no longer be. Bogost's cool yet captivating prose style succeeds in seducing us into a blissful recognition of the wondrous strangeness of commonplace objects. To object to this vision makes one feel a bit like an environmentalist version of the feminist killjoy.[64] Who could resist this lusty invitation: "The density of being makes it *promiscuous*, always touching everything else, unconcerned with differentiation. Anything is enough to party."[65] This pro-sex feminist cannot help but be enticed by such an invitation, yet I suspect that consumerist orgies produce particularly toxic hangovers. Fortunately, Annie Sprinkle and Beth Stephens, in their "Ecosexual Manifesto,"[66] invite us to more exuberant, less capitalist festivities. Such pleasures, from the ethical modes of inhabiting with which this book began, to the agonizing ecodelic anthropocene dissolves, refuse the prevailing disconnection of the human from everything and everyone else. Thinking as the stuff of the world entails thinking in place, in places that are simultaneously the material of the self and the vast networks of material worlds.

The supposedly awesome plentitude of objects looks different from the perspectives of postcolonial ecologies, environmental health and environmental justice movements, those enduring "slow violence," climate refugees, Bhopal survivors, ocean conservation movements, and from the perspectives of myriad nonhuman creatures. We could wonder what it would be like to "be" a plastic bag or a plastic bottle cap.

Or we could consider the networks of chemistry, capitalist consumerism, inland waterways, ocean currents, and addiction to high-fructose corn syrup that have created the Great Pacific Garbage patch. Rather than seeing the world as composed of inert resources or as discrete objects, new materialist theories and new political movements, focused on stuff, stress the strange agencies of everyday things. The recognition that banal objects such as toothbrushes, razors, plastic bottles, plastic bags, food containers, children's toys, and so forth, intended for momentary human use, pollute for eternity, renders them weirdly malevolent. As chapter 5 discussed, activists against plastic pollution dramatize the strange agencies of plastics, which absorb toxins in the seas, enter the ocean food chain, and harm birds, mammals, reptiles, and fishes, causing disease, obstructing airways, or clogging digestive tracks. The artist Pam Longobardi, who collects plastic ocean debris and uses it as her material, explains how the everyday objects that surround terrestrial humans become something quite different when understood from the perspective of ocean ecologies: "The plastic elements initially seem attractive and innocuous, like toys, some with an eerie familiarity and some totally alien. At first, the plastic seems innocent and fun, but it is not. It is dangerous. In our eagerness for the new, we are remaking the world in plastic, in our own image, this toxic legacy, this surrogate, this imposter."[67]

The plastic anthropocene, manufactured by humans but beyond human control, not only surrounds us but invades us, literally transforming our flesh as Tuana explains. Neither the predominant paradigm of sustainability nor the transcendent anthropocene visions discussed in the prior chapter can account for the human as permeable flesh. Nor can they reckon with incommensurable grids of vulnerability, culpability, responsibility, and concern. We need environmentalisms that refuse to take refuge within ontoepistemologies that are not tenable within a world of rapid anthropogenic alterations, strange agencies, and precarious human and nonhuman lives. The practice of thinking from within and as part of the material world swirls together ontology, epistemology, scientific disclosures, political perspectives, posthuman ethics, and environmental activism. There is no position outside, no straight path, no belief in transparent global systems of knowledge, only modest protests and precarious pleasures, from within compromised locations shadowed by futures that will surely need repair.

Acknowledgments

◇◇◇

This book, even more than most, was catalyzed by other scholars, since nearly every chapter was sparked by an invitation to speak at an event or write for a collection. The much-revised chapters benefited from many people's comments as well as ongoing scholarly conversations—face to face, at events, and through social media. I sincerely regret that it is impossible to thank everyone here. Not only is my memory just wretched, but you rarely catch the name of the person who asked that provocative question at your talk, and even if you do jot it down, where does that bit of paper end up? In a time when the academy, and particularly research in the humanities, is under assault, it is vital, however, to acknowledge communities of passionate scholars committed to work that attempts to matter.

Therefore, I would like to extend my gratitude to the following people and institutions whose invitations initiated these essays: Gregory Caicco, the Lincoln Center for Ethics, and Arizona State University, for the invitation to present at the symposium "Ethics in Place: Architecture, Memory, and Environmental Ethics"; Catriona Sandilands, Bruce Erickson, and York University, who invited me to participate in the workshop "Queer Ecologies: Sex, Nature, Biopolitics, and Desire"; Hilda Rømer Christensen, Michala Hvidt Breengaard, Helene Hjorth Oldrup, and the Coordination for Gender Research, University of Copenhagen, for the invitation to speak at the "Gendering Climate and Sustainability" conference; Joni Adamson, Sally Kitsch, and the Institute for Humanities Research at Arizona State University, for the invitation to participate in the "Symposium on Humanities and Human Origins"; Heather Sullivan and Dana Phillips, for inviting me to contribute to a special issue of *ISLE* on "Material Ecocriticism"; Serpil Opperman and Serenella Iovino, for inviting me to contribute to their edited collection *Material Ecocriticism*; Richard Grusin, John C. Blum, Emily Clark, and Dehlia Hannah for the invitation to speak at the C21 conference "Anthropocene Feminism" at the University of Wisconsin–Milwaukee; Eileen Joy, for inviting me to contribute to the inaugural edition of *O-Zone: A Journal of Object Oriented Studies*; and the late Patricia Yaeger, for asking me to contribute the PMLA "Theory and Methodology" section on "Sustainability." I also presented versions of these chapters as invited talks, plenary presentations, and workshops and am deeply grateful to the generous invitations, the lively discussions, and the collegiality shown by so many at all of those events. In particular, though, I would like to thank (the incomparable) Eileen Joy, the Babel Working Group, and UC Santa Barbara, for asking me to speak at "BABEL: On the Beach" (where we were, gloriously, on the beach and in the ocean!); Jeffrey J. Cohen and GW MEMSI at George Washington University, for the invitation to participate in the "Transition, Scale, and Catastrophe" symposium; Serpil Opperman, Micha Castellanos, Hanna Strass, and EASLCE (European Association for the Study of Literature, Culture, and the Environment), for the invitation to conduct a webinar about my work; Angie Bennett Segler, for the Google+ Hangouts discussion on "Embodiment and Materiality"; Anna Banks, Erin James, Jennifer Ladino, Jodie Nicotra, and Scott Slovic, for inviting me to speak at the University of Idaho; Cecilia Äsberg and Magda Górska, for the invitation to be the plenary speaker for the European "New Materialism" conference at University of Linköping, Sweden; Denise Buell, for inviting me to speak at Williams College; the members of the "Materialisms and New Materialisms across the Disciplines" seminar at Rice University, for the invitation to present and conduct a workshop; the environmental studies and English department faculty at University of North Texas, including Priscilla Ybarra, for the invitation to present; the "Gender, Science, and Critical Animal Studies" conference at SUNY Stonybrook; the "Gender, Violence and Agency in an Era of Globalization Initiative," at the University of Vienna; and David Vázquez, Richard York, Kari Norgaard, Carol Stabile, and Karen Ford at the University of Oregon, for invitations to speak. (Although I gave many other presentations during this period, those talks dovetail with my next book, so those acknowledgments will appear there.)

I was honored to teach the European doctoral seminar "Introducing Feminist Materialisms" for Inter-Gender, the Swedish International Research School in Interdisciplinary Gender Studies, with Nina Lykke and Cecilia Äsberg at the University of Linköping; tack så mycket for that invigorating experience! I am also grateful to the graduate students at the University of Texas at Arlington who participated in my seminars on "New Materialisms," "Science Fiction and Posthumanism in the Anthropocene," "The Human after the Nonhuman Turn," and "Gender, Race, and Sexuality in the Wake of Social Construction."

Thank you to Dan Phillipon, Bruce Braun, Doug Armato, Richard Morrison, the Quadrant Project, the University of Minnesota Press, and the University of Minnesota Institute for Advanced Study for the invitation to speak and spend time with the members of the faculty and the press. I was struck by the enthusiasm and intellectual capaciousness of the University of Minnesota Press and am grateful to Doug Armato for his continued interest in my work, to Dani Kasprzak for helping me map out the revisions, and to Anne Carter for keeping the editorial chaos at bay. (Anne's enthusiasm for the hedgehogs was also quite appreciated!) I am wild about the cover design and thank the editorial and design team for their insistence that we come up with something playful—particular thanks to Jeenee Lee, the designer. Nicholas Taylor caught many inconsistencies during copyediting, and Mike Stoffel provided solid direction. I am very grateful to the two external readers for this project, both of whom identified themselves—Jeffrey J. Cohen and Jack Halberstam. The revisions greatly benefited from their insightful and provocative critiques as well as their generous suggestions for additional scholarship to consider. While I was unable to follow all their leads, the book is significantly better because of them.

I am grateful to Linda Hogan and Coffee House Press, for allowing me to quote from her poetry. I am also quite grateful to the artists, organizations, and activists who were willing to share their work here: Kirsten Justesen, Katrin Peters, Greenpeace Switzerland and Spencer Tunick, Mike Grenville and Bare Witness, and David Thomas Smith. Special thanks to the following artists (and a scientist) for their extraordinary generosity: Patricia Johanson, Pam Longobardi, John Megahan, and Tim Senden of the Australian National CT Lab. And I am thrilled to have a still image from Marina Zurkow's video "Slurb" on the cover—thank you, Marina. I would also like to thank the presses and journals that allowed me to publish revised versions of earlier essays: University Press of New England, Indiana University Press, University of Minnesota Press, *Women, Gender and Research* (*Kvinder, Køn og Forskning*, Denmark), *O-Zone: A Journal of Object-Oriented Studies, ISLE, PMLA*, and *Women and Performance* (as well as the anonymous reviewers for that journal whose critiques I only managed to address in this final version of the essay).

Not only does the work of Catriona Sandilands, Ursula Heise, and Jeffrey J. Cohen inspire me, but I value their friendship, intellectual provocations, and support. I am quite grateful for the generous support of Larry Buell and Cary Wolfe. My thinking on matters pertaining to this volume has also been enriched by conversations (live or virtual) with the following extraordinary scholars (just recently or over the years): Joni Adamson, Cecilia Äsberg, Karen Barad, Bridgitte Barclay, Jamie Skye Bianco, Suzanne Bost, Rosi Braidotti, Ron Broglio, John Bruni, Levi Paul Bryant, Larry Buell, Allison Carruth, Dianne Chisholm, Jon Christensen, Lucinda Cole, Claire Colebrook, Elizabeth DeLoughrey, Giovanna DiChiro, Adam Dickinson, Lowell Duckert, Stephanie Foote, Samantha Frost, Greta Gaard, Greg Garrard, Magdalena Górska, Richard Grusin, Donna J. Haraway, Stephen Harding, Eva Hayward, Susan Hekman, Lissa Hollaway-Attaway, Heather Houser, Serenella Iovino, Jennifer James, Eileen Joy, Melody Jue, Katie King, Stephanie LeMenager, Anthony Lioi, Bonnie Mann, Bob Markley, Robert McRuer, Ladelle McWhorter, Steve Mentz, Astrida Neimanis, Rob Nixon, Sharon O'Dair, Serpil Opperman, Paul Outka, Dana Phillips, Karen Raber, Sangeeta Ray, Sarah Jaquette Ray, Celia Roberts, Siobhan Senier, Rebekah Sheldon, Hanna Sjögren, Susan Squier, Nicole Starosielski, Karl T. Steel, Rajani Sudan, Heather Sullivan, Julie Sze, Christy Tidwell, Nancy Tuana, Iris van der Tuin, Jamie Weinstein, Cary Wolfe, and Marina Zurkow. Warm thanks to Vanessa Daws, the Irish swim artist, for inviting me to swim in and write for one of her events.

I am fortunate to have had such wonderful colleagues, including graduate students and former graduate students at the University of Texas at Arlington: Barbara Chiarello, Tra Clough, Sean Farrell, Jackie Fay, Luanne Frank, Jim Grover, Kevin Gustafson, Audrey Haferkamp, Steve Harding, Susan Hekman, Desiree Henderson, Penny Ingram, Douglas Klahr, Laura Kopchik, Bruce Krajewski, Robert LaRue, Justin Lerberg, Matthew Lerberg, Jeffrey Marchand, Neill Matheson, Cedrick May, Julie McCown, Kandice San Miguel, Chris Morris, Tim Morris, Laura Mydlarz, Kevin Porter, Tim Richardson, Ken Roemer, Rod Sachs, Peggy Semingson, Sarah Shelton, Johanna Smith, Connor Stratman, Meghna Tare, Stephanie Peebles Tavera, Amy Tigner, Jim Warren, Kathryn Warren, Hue Woodson, and the late Ben Agger. The generous support of my former department chair, Wendy Faris, has been invaluable. My dearest pals in Dallas make everything better, always. A toast to you, Rajani Sudan, Chris Morris, Karen Coates, Susan Hekman, Tom Pribyl, Wendy Faris, Neill Matheson, and Colleen Fitzgerald! Paula Frasier deserves special thanks for knowing exactly how to take care of everything and everyone. I am grateful to many friends and family members for their kindness and support: deepest gratitude to my sister, Kathy Hoglund; my mother, Charlotte Fisk; my stepmother, Mary Brown; and my dearest friend, Tracy Sweetland. Many thanks to Evan Engwall, a most flexible and wise coparent.

How fabulous it is to spend my life with Emma Alaimo and Kai Engwall, who are kind, curious, thoughtful, funny, and adventurous—ready to snorkel with gators and manatees, swim through a cave, or climb to the top of a ruin in Tikal. And I am so lucky as well to have (or have had) the daily amusements and affections of Carmel, Pip, Felix, and Crackers.

What bliss to end the acknowledgments with Stephanie LeMenager, as we are only just beginning. Her brilliance, wit, and exuberance are sheer delight. I dedicate *Exposed* (formerly entitled *Protest and Pleasure*) to her, who was once a Lesbian Avenger.

Notes

Introduction

1. Rosi Braidotti, *Transpositions* (Malden, Mass.: Polity, 2006), 278.
2. Stacy Alaimo, *Bodily Natures: Science, Environment, and the Material Self* (Bloomington: Indiana University Press, 2010). I realize hope is a powerful political force, and for many people indispensable for survival, but I prefer the Buddhist ideal of "right intention," performed while detaching from projected outcomes. This sense of relinquishing control, while perhaps problematic for perpetual political struggles which must project forward in time, makes sense for a new materialist understanding of emergence and intra-active agencies.
3. Braidotti, *Transpositions*, 278.
4. Ursula K. Heise, *Sense of Place and Sense of Planet: The Environmental Imagination of the Global* (New York: Oxford University Press, 2008), 62.
5. Ibid., 210.
6. Ulrich Beck, *Risk Society: Towards a New Modernity*, trans. Mark Ritter (London: Sage, 1992), 54.
7. Adriana Petryna, *Life Exposed: Biological Citizens after Chernobyl* (Princeton, N.J.: Princeton University Press, 2002), 216.
8. See, for example, the conference "The Pedagogics of Unlearning," organized by Eamonn Dunne, Aiden Seery, and Michael O'Rourke; the "Irrationale" posted on the conference website states, "To maintain a fidelity to learning is to be faithful to the call of stupidity, to adopt a position of epistemological and pedagogical humility, to be humble before the other, just because one simply doesn't know." See http://www.unlearningconf.com/irrationale/.
9. The sixth extinction, also termed the holocene or anthropocene extinction, does not have a clear starting point, but it is obvious that the rate of extinction has accelerated into the twenty-first century and that these extinctions, by and large, are anthropogenic. Here is one way of approaching its enormity: one million species, or between 30 and 50 percent, of all species are expected to become extinct by 2050.
10. Eileen Joy, in Eileen Joy and L. O. Aranye Fradenburg, "Unlearning: A Dialogue," unpublished text from talk for "The Pedagogics of Unlearning" (Trinity College, Dublin, September 6–7, 2014), 16, 13.
11. Judith Halberstam, *The Queer Art of Failure* (Durham, N.C.: Duke University Press, 2011), 24.
12. Braidotti, *Transpositions*, 278.

13. Laura Kipnis, "(Male) Desire and (Female) Disgust: Reading Hustler," in *Cultural Studies*, ed. Lawrence Grossberg, Cary Nelson, and Paula Treichler (New York: Routledge, 1992), 376, 379.

14. Halberstam, *Queer Art of Failure*, 16.

15. See my discussions of feminist epistemology in *Undomesticated Ground: Recasting Nature as Feminist Space* (Ithaca, N.Y.: Cornell University Press, 2000); and Sandra Harding, *Sciences from Below: Feminisms, Postcolonialisms, Modernities* (Durham, N.C.: Duke University Press, 2008).

16. Donna J. Haraway, "Situated Knowledges: The Science Question in Feminism and the Privilege of Partial Perspective," in *Simians, Cyborgs, Women: The Reinvention of Nature* (New York: Routledge, 1991), 188, 189.

17. Ibid., 190.

18. Stacy Alaimo and Susan Hekman, eds., *Material Feminisms* (Bloomington: Indiana University Press, 2009).

19. Karen Barad, *Meeting the Universe Halfway* (Durham, N.C.: Duke University Press, 2007), 185.

20. Ibid.

21. Jeffrey Jerome Cohen, "Introduction," in *Prismatic Ecologies: Ecotheory beyond Green*, ed. Cohen (Minneapolis: University of Minnesota Press, 2013), xxiv.

22. Bruce Braun and Sarah J. Whatmore, "The Stuff of Politics: An Introduction," in *Political Matter: Technoscience, Democracy and Public Life*, ed. Braun and Whatmore (Minneapolis: University of Minnesota Press, 2010), xi.

23. Mick Smith, *Against Ecological Sovereignty: Ethics, Biopolitics, and Saving the Natural World* (Minneapolis: University of Minnesota Press, 2011), 183.

24. Barad, *Meeting the Universe Halfway*, 185.

25. Cohen, "Introduction," in *Stone: An Ecology of the Inhuman* (Minneapolis: University of Minnesota Press, 2015), 9.

26. Stephanie LeMenager, *Living Oil: Petroleum Culture in the American Century* (New York: Oxford University Press, 2014), 6. LeMenager is drawing on the work of Philip Auslander to describe the contemporary mediatization of lived experience. The point is more nuanced in her account.

27. For a different perspective, which argues for a new ecofeminism, see Greta Gaard's well-researched, comprehensive essay, which argues that a broad array of issues should be analyzed from the perspective of ecological feminism, including "global gender justice; climate justice; sustainable agriculture; healthy and affordable housing; universal and reliable health care, particularly maternal and infant health care; safe, reliable, and free or low-cost reproductive technologies; food security; sexual self determination; energy justice; interspecies justice; ecological, diverse, and inclusive educational curricula; religious freedom from fundamentalisms; indigenous rights; the production and disposal of hazardous wastes; and more." Gaard argues that these quite different issues could be brought

within an ecofeminist analytics: "An intersectional ecological–feminist approach frames these issues in such a way that people can recognize common cause across the boundaries of race, class, gender, sexuality, species, age, ability, nature—and affords a basis for engaged theory, education, and activism." Greta Gaard, "Ecofeminism Revisited: Rejecting Essentialism and Re-placing Species in a Material Feminist Environmentalism," *Feminist Formations* 23, no. 2 (2011): 44.

28. Richard Twine, in 2001, argued that "ecofeminism can consolidate its tradition of elucidating the interconnections between different oppressions by expanding upon its philosophy of the body," and he stresses that "those ideas which encourage us to devalue certain bodies stem from discourses related to nature and animality." See Twine, "Ma(r)king Essence: Ecofeminism and Embodiment," *Ethics and Environment* 6, no. 2 (2001): 31. His argument about how new conceptualizations of essence can disrupt dualisms has much in common with material feminisms and preceded the collection *Material Feminisms* by nearly a decade.

29. See the prior note, on Richard Twine's argument about essentialism.

1. This Is about Pleasure

1. Gregory Caicco, "Introduction," in *Architecture, Ethics, and the Personhood of Place*, ed. Gregory Caicco (Lebanon, N.H.: University Press of New England, 2007), 1.

2. See Amy Kaplan's "Manifest Domesticity," in *No More Spheres: A Next Wave American Studies Reader*, ed. Cathy N. Davidson and Jessamyn Hatcher (Durham, N.C.: Duke University Press, 2002), 184–207, which argues that in antebellum America the ideology of domesticity was actually linked to that of manifest destiny; they were not, in fact, separate spheres, politically speaking.

3. Gaston Bachelard, *The Poetics of Space*, trans. Maria Jolas (Boston: Beacon Press, 1964), 6.

4. Nina Baym, *Woman's Fiction: A Guide to Novels by and about Women in America, 1820–1870* (Ithaca, N.Y.: Cornell University Press, 1978).

5. Stacy Alaimo, *Undomesticated Ground: Recasting Nature as Feminist Space* (Ithaca, N.Y.: Cornell University Press, 2000).

6. For an analysis of Austin's Walking Woman, see ibid., chapter 3.

7. Recent victories for marriage equality notwithstanding here, since bringing queer people into the marital fold is a form of domestication, in at least two senses of that word.

8. Donna Haraway, *The Companion Species Manifesto: Dogs, People, and Significant Otherness* (Chicago: Prickly Paradigm Press, 2003); Michael Pollan, *The Botany of Desire: A Plant's Eye View of the World* (New York: Random House, 2002).

9. Yi-Fu Tuan, *Dominance and Affection: The Making of Pets* (New Haven, Conn.: Yale University Press, 1984), 99.

10. Donna Haraway, *Simians, Cyborgs, and Women: The Reinvention of Nature* (New York: Routledge, 1991), 181; Carolyn Merchant, *Earthcare: Women and the Environment* (New York: Routledge, 1995), 217; Val Plumwood, *Feminism and the Mastery of Nature* (New York: Routledge, 1993), 124.

11. Jacob Bronowski, as quoted in Jerry Kastner and Brian Wallis, *Land and Environmental Art* (London: Phaidon, 1998), 11.

12. A striking contrast here would be Japanese architecture, in which "the absence of a well defined center and the ambiguous spatial quality . . . are predicated on both the lack of a ruling order, and the layered, indefinite boundaries." The Japanese house, then, is a space in which "everything—forms, entities, together with their meanings[—is] continuously open-ended or unfinished." Botond Bognar, "The Place of No-Thingness: The Japanese House and the Oriental World Views of the Japanese," in *Dwellings, Settlements, and Traditions: Cross-Cultural Perspectives*, ed. Jean-Paul Bourdier and Nezar AlSayyad (New York: Lanham, 1989), 196.

13. Alberto Pérez-Gómez and Louise Pelletier, *Architectural Representation and the Perspectival Hinge* (Cambridge, Mass.: MIT Press, 1997), 384.

14. Walter de Maria, New York Earth Room, 1977. Installed at the Dia Center for the Arts, New York.

15. Luce Irigaray, *This Sex Which Is Not One*, trans. Catherine Porter (Ithaca, N.Y.: Cornell University Press, 1985), 77.

16. Nan Ellin, "Shelter from the Storm or Form Follows Fear and Vice Versa," in *The Architecture of Fear*, ed. Nan Ellin (New York: Princeton Architectural Press, 1997), 33, 42.

17. Edward J. Blakely and Mary Gail Snyder, "Divided We Fall: Gated and Walled Communities in the United States," in Ellin, *Architecture of Fear*, 85.

18. Lisa Lewenz, "Three Mile Island Calendar," 1984. Mass produced in Chicago. Digital image available at University of Oregon library: http://oregondigital.org/catalog/oregondigital:df70bt88v.

19. Todd Haynes, liner notes for *Safe*, dir. Todd Haynes (Chemical Films Limited Partnership, 1995).

20. For an extended reading of *Safe* in terms of the material agencies of environmental illness / multiple chemical sensitivity, see Stacy Alaimo, *Bodily Natures: Science, Environment, and the Material Self* (Bloomington: Indiana University Press, 2010).

21. Simon C. Estok, "Painful Material Realities, Tragedy, Ecophobia," in *Material Ecocriticism*, ed. Serenella Iovino and Serpil Oppermann (Bloomington: Indiana University Press, 2015), 130.

22. I was introduced to living architecture in a lecture Allison Carruth gave at the University of Texas at Arlington. For her discussion of Oliver

Medvedik's In Vitro Meat Habitat, and other plans for biological architecture, see Carruth, "The City Refigured: Environmental Vision in a Transgenic Age," in *Environmental Criticism for the 21st Century*, ed. Stephanie LeMenager, Teresa Shewry, and Ken Hiltner (New York: Routledge, 2012), 85–104.

23. Linda Hogan, "The History of Red," in *The Book of Medicines* (Minneapolis: Coffee House Books, 1993), 9.

24. Jodi A. Byrd, *The Transit of Empire: Indigenous Critiques of Colonialism* (Minneapolis: University of Minnesota Press, 2011), 118.

25. Hogan, "History of Red," 9.

26. Ibid., 11.

27. Linda Hogan, "The Bricks," in *Book of Medicines*, 67.

28. Ibid.

29. This image haunts Hogan's prose as well: "In Japan, I recall, there were wildflowers that grew in the far, cool regions of mountains. The bricks of Hiroshima, down below, were formed of clay from these mountains, and so the walls of houses and shops held dormant trumpet flower seeds. But after one group of humans killed another with the explosive power of life's smallest elements split wide apart, the mountain flowers began to grow. Out of the crumbled, burned buildings they sprouted. Out of destruction and bomb heat and the falling of walls, the seeds opened up and grew. What a horrible beauty, the world going its own way, growing without us." Linda Hogan, *Dwellings: A Spiritual History of the Living World* (New York: Simon and Schuster, 1995), 33.

30. Hogan, "The Bricks," 68.

31. Ibid.

32. Ibid.

33. Katherine McKittrick, *Demonic Grounds: Black Women and the Cartographies of Struggle* (Minneapolis: University of Minnesota Press, 2006), 2.

34. Octavia Butler, *Imago* (New York: Warner Books, 1989), 33.

35. For "rendering," see, of course, Nicole Shukin, *Animal Capital: Rendering Life in Biopolitical Times* (Minneapolis: University of Minnesota Press, 2009).

36. McKittrick, *Demonic Grounds*, 141.

37. *Habitat*, dir. Rene Daalder (Matarans, 1998). For a more extensive analysis of both *Safe* and *Habitat*, see Stacy Alaimo, "Discomforting Creatures: Monstrous Natures in Recent Films," in *Beyond Nature Writing: Expanding the Boundaries of Ecocriticism*, ed. Karla Armbruster and Kathleen R. Wallace (Charlottesville: University Press of Virginia, 2001), 279–311.

38. I would like to think the fictional Hank was inspired by Juhani Pallasmaa's *Eyes of the Skin*, in which he states that architecture is "essentially an extension of nature into the manmade realm, providing the ground for perception and the horizon to experience and understand the world."

See Pallasmaa, *The Eyes of the Skin: Architecture and the Senses* (London: Academy Group, 1996), 28.

39. J. K. Gibson-Graham, *The End of Capitalism (As We Knew It): A Feminist Critique of Political Economy* (Minneapolis: University of Minnesota Press, 2006), x.

40. Stephanie LeMenager, *Living Oil: Petroleum Culture in the American Century* (New York: Oxford University Press, 2014), 75, 74.

41. Ibid., 75.

42. Catriona Sandilands, "Sex at the Limits," in *Discourses of the Environment*, ed. Eric Darier (Oxford, U.K.: Blackwell, 1999), 79.

43. Dan Phillipon, "Sustainability and the Humanities: An Extensive Pleasure," *American Literary History* 24, no. 1 (2012): 173, 172, 171.

44. See "Meet Ron Finley," n.d., http://ronfinley.com/meet-ron-finley/; and "Ron Finley: A Guerilla Gardener in South Central L.A," TED talk, February 2013, http://www.ted.com/talks/ron_finley_a_guerilla_gardener _in_south_central_la?language=en.

45. Gail Weiss, *Body Images: Embodiment as Intercorporeality* (New York: Routledge 1998), 158.

46. Moira Gatens, *Imaginary Bodies: Ethics, Power, and Corporeality* (New York: Routledge, 1996), 110.

47. Julia Butterfly Hill, *The Legacy of Luna: The Story of a Tree, a Woman, and the Struggle to Save the Redwood* (San Francisco: Harper, 2000), 227.

48. James Ficklin, dir., *Tree-Sit: The Art of Resistance* (Earth Films, 2001).

49. Weiss, *Body Images*, 5.

50. Ladelle McWhorter, *Bodies and Pleasures: Foucault and the Politics of Sexual Normalization* (Bloomington: Indiana University Press, 1999), 167.

51. Ibid., 192.

52. Catriona Sandilands, "Desiring Nature, Queering Ethics: Adventures in Erotogenic Environments," *Environmental Ethics* 23, no. 2 (2001): 188.

53. Mary Oliver, "Wild Geese," in *New and Selected Poems* (Boston: Beacon Press, 1993).

54. See also, of course, William Cronon, "The Trouble with Wilderness; or, Getting Back to the Wrong Nature," in *Uncommon Ground: Rethinking the Human Place in Nature*, ed. William Cronon (New York: Norton, 1996), 69–90.

55. Rebecca Solnit, *As Eve Said to the Serpent: On Landscape, Gender, and Art* (Athens: University of Georgia Press, 2001), 164.

56. Edward Casey, *Getting Back into Place: Toward a Renewed Understanding of the Place-World* (Bloomington: Indiana University Press, 1993), 255, 256.

57. Since the early version of this chapter was originally published, green architecture, sustainable architecture, and other styles of environmental architecture have rapidly proliferated—so much so that it would be

impossible to provide any sort of comprehensive account here. None-theless, the chapter suggests how thinking with architecture informed the development of my new materialist theory of trans-corporeality. Many people have asked me to discuss whether positive, pleasurable versions of trans-corporeality are possible. "This Is about Pleasure" does advocate positive modes of posthumanist, multi-species intercor-poreality, which moves in the direction of trans-corporeality. I would distinguish trans-corporeality from the pleasurable multi-species ver-sion of intercorporeality I posit here, in that trans-corporeality offers a coherent conception of subjectivity as material as well as scientifically and politically mediated. Trans-corporeality emphasizes the necessity for scientific "captures" of material agencies. Furthermore, my concep-tion of the trans-corporeal subject draws on the theories of Judith But-ler, Donna Haraway, Andrew Pickering, Ulrich Beck, and Karen Barad, which this chapter does not manifest. I welcome disagreement on these distinctions, however, as many people have taken up and transformed my concept of trans-corporeality.

58. Sherry Ahrentzen, "The Space between the Studs: Feminism and Archi-tecture," *SIGNS* 29, no 1 (2003): 179–206.

59. Margaret A. Somerville, "Ethics and Architects: Spaces, Voids, and Travelling-in-Hope," in *Architecture, Ethics, and Technology*, ed. Louise Pel-letier and Alberto Pérez-Gómez (Montreal: McGill-Queens University Press, 1994), 72, 77.

60. Karsten Harries, *The Ethical Function of Architecture* (Cambridge, Mass.: MIT Press, 2002), 180.

61. George Dodds and Robert Tavernor, eds., *Body and Building: Essays on the Changing Relation of Body and Architecture* (Cambridge, Mass.: MIT Press, 2002); Kent C. Blooomer and Charles W. Moore, *Body, Memory, and Archi-tecture* (New Haven, Conn.: Yale University Press, 1977).

62. Deborah Fausch, "The Knowledge of the Body and the Presence of His-tory: Toward a Feminist Architecture," in *Architecture and Feminism*, ed. Debra Coleman, Elizabeth Danze, and Carol Henderson (New York: Princeton Architectural Press, 1996), 40.

63. Elizabeth Grosz, *Architecture from the Outside: Essays on Virtual and Real Space* (Cambridge, Mass.: MIT Press, 2001), 99, 100.

64. Catriona Sandilands, *Good-Natured Feminist: Ecofeminism and the Quest for Democracy* (Minneapolis: University of Minnesota Press, 1999), 181.

65. Sonfist, as quoted in Kastner and Wallis, *Land and Environmental Art*, 258.

66. Ursula K. Heise makes a similar point in her analysis of cyborg animals: "What underlies the imaginative exploration of artificial animals, then, is the question of how much nature we can do without, to what extent simulations of nature can replace the 'natural,' and what role animals,

both natural and artificial, play in our self-definition as humans." See Heise, "From Extinction to Electronics: Dead Frogs, Live Dinosaurs, and Electric Sheep," in *Zoontologies: The Question of the Animal*, ed. Cary Wolfe (Minneapolis: University of Minnesota Press, 2003), 60.

67. Jennifer Wolch, "Zoöpolis," *Capitalism, Nature, Socialism: A Journal of Socialist Ecology* 7, no. 26 (1996): 29.

68. Lance Richardson, "New York Needs Coyotes," *Slate*, July 31, 2015.

69. Quoted in ibid.

70. *Winged Migration (Le Peuple Migrateur)*, dir. Jacques Perrin (Sony, 2001).

71. Ibid.

72. Claire Colebrook, "From Radical Representations to Corporeal Becomings: The Feminist Philosophy of Lloyd, Grosz, and Gatens," *Hypatia* 15, no. 2 (2000): 76–93.

73. Mick Smith, *Against Ecological Sovereignty: Ethics, Biopolitics, and Saving the Natural World* (Minneapolis: University of Minnesota Press, 2011), 97.

74. Johanson, quoted in Caffyn Kelley, *Art and Survival: Patricia Johanson's Environmental Projects* (Salt Spring Island, B.C.: Islands Institute of Environmental Studies, 2006), 21.

75. Xin Wu, *Patricia Johanson and the Re-invention of Public Environmental Art, 1958–2010* (Surrey, U.K.: Ashgate, 2013), 145.

76. Ibid., 148.

77. Johanson, quoted in Kelley, *Art and Survival*, 19.

78. Johanson, quoted in ibid., 26. One of Johanson's more provocative proposals is the "Garden of Organized Killing / Soil Fertility" project, which would be positioned outside a slaughterhouse, where "the horror of organized killing is brought into the open in a public garden that enriches the soil with the blood of sacrificial victims. The stench of death is everywhere." Johanson quoted in ibid., 26.

79. Goto, as quoted in Baile Oakes, *Sculpting with the Environment: A Natural Dialogue* (New York: Van Nostrand-Reinhold, 1995), 135.

80. Artist statement, n.d., http://collinsandgoto.com/artwork/.

81. Hull, quoted in Oakes, *Sculpting with the Environment*, 139.

82. Hull, "Bio," n.d., http://eco-art.org/?page_id=7.

83. Hull, "Proposals Archive and Active," n.d., http://eco-art.org/?page_id=32.

84. Hull, "Get Involved," n.d., http://eco-art.org/?page_id=17.

85. Brian Massumi, *Parables of the Virtual: Movement, Affect, Sensation* (Durham, N.C.: Duke University Press, 2002), 236.

86. Grosz, *Architecture from the Outside*, 104.

87. Jeffrey Jerome Cohen, "Introduction," in *Prismatic Ecologies: Ecotheory beyond Green*, ed. Cohen (Minneapolis: University of Minnesota Press, 2013), xxv.

2. Eluding Capture

1. Dana Seitler documents the emergence of sexual "perversity" as interconnected with other categories: "The construction of perversity appears as part of a story in which race, gender, physical deformation, sexuality, and many other bodily forms and practices emerge in ontologically and epistemologically interdependent ways." Seitler, "Queer Physiognomies; or, How Many Ways Can We Do the History of Sexuality?" *Criticism* 46 (Winter 2004): 74. See also the short pieces by many scholars within the "Theorizing Queer Inhumanisms" "Dossier," which includes many essays on race and indigeneity. "Theorizing Queer Inhumanisms," *GLQ* 21, nos. 2–3 (2015): 209–48.

2. Catriona Sandilands, "Queer Life? Ecocriticism after the Fire," in *The Oxford Handbook of Ecocriticism*, ed. Greg Garrard (New York: Oxford University Press, 2015), 315.

3. Karen Barad, "Transmaterialities: Trans/Matter/Realities and Queer Political Imaginings" *GLQ* 21, nos. 2–3 (2015): 387.

4. Bruce Bagemihl, *Biological Exuberance: Animal Homosexuality and Natural Diversity* (New York: St. Martin's Press, 1999), 265.

5. Ibid., 1–2.

6. Joan Roughgarden, *Evolution's Rainbow: Diversity, Gender, and Sexuality in Nature and People* (Berkeley: University of California Press, 2004).

7. "Against Nature? An Exhibition on Animal Homosexuality," Naturhistorisk museum, Oslo, Norway, 2007, http://www.nhm.uio.no/besokende/skiftende-utstillinger/againstnature.

8. Donna Haraway, *Primate Visions: Gender, Race, and Nature in the World of Modern Science* (New York: Routledge, 1989), 26.

9. "Gay Outrage over Penguin Sex Test," *BBC News*, February 14, 2005, http://news.bbc.co.uk/2/hi/europe/4264913.stm.

10. "Penguins Can Stay Gay," *Ananova*, 2005, http://www.ananova.com/news/story/sm_1284769.

11. Olivia Judson, *Dr. Tatiana's Sex Advice to All Creation: The Definitive Guide to the Evolutionary Biology of Sex* (New York: Vintage, 2003), 143.

12. Susan Block, "The Bonobo Way," n.d., http://www.blockbonobo foundation.org/. Block is not the only one inspired by Bonobo sex. Barbara Ehrenreich, in a piece titled "Let Me Be a Bonobo," predicts a "surge in trans-species people, who will eagerly go over to the side of the chimps," suggesting that one "reason to make the human-to-ape transition is the sex": "Bonobos, genetically as close to humans as larger chimpanzees, use sex much as we use handshakes—as a form of greeting between individuals in any gender combination." See Ehrenreich, "Let Me Be a Bonobo," *The Guardian*, May 10, 2007. Kelpie Wilson's science fiction novel *Primal Tears* features a half-bonobo, half-human protagonist named "Sage." See

204 ᐳᐳ NOTES TO CHAPTER 2

Wilson, *Primal Tears* (Berkeley, Calif.: Frog Ltd., 2005). Interestingly, the same sort of alliance between sexual freedom and environmentalism that Susan Block promotes becomes a problem in the novel when some of Sage's fans transform her "Rainbow Clubs"—which are intended to promote the protection of endangered bonobos—into sex clubs.

13. "Against Nature?"

14. Roughgarden, *Evolution's Rainbow*, 128.

15. Bagemihl, *Biological Exuberance*, 107.

16. Margaret Cuonzo, "Queer Nature, Circular Science," in *Science and Other Cultures: Issues in Philosophy of Science and Technology*, ed. Robert Figueroa and Sandra Harding (New York: Routledge, 2003), 231.

17. Cuonzo also refers to the "other minds' problem," questioning whether, say, the illustrations in Bagemihl's book, "pictures of animals in what looks like sexual activity," are, in fact, sex: "But how do we know that these behaviors are what they seem to be?" Ibid., 230. While it is epistemologically and ethically useful to underscore the limits of human knowledge, it is just as problematic to conclude that because we cannot, absolutely, know these behaviors "are" sex, then they must not be. Certainly, heterosex between animals is not held up to such a high standard of "proof." Cuonzo's skepticism seems a perfect example of how cultural critics are much better (in Latour's terms) at "subtracting reality."

18. Catriona Sandilands, "Unnatural Passions? Notes toward a Queer Ecology," *Invisible Culture: An Electronic Journal for Visual Culture* 9 (Fall 2005): n.p.

19. Giovanna Di Chiro, "Polluted Politics? Confronting Toxic Discourse, Sex Panic, and Econormativity," in *Queer Ecologies*, ed. Catriona Mortimer-Sandilands and Bruce Erickson (Bloomington: Indiana University Press, 2010), 199–230.

20. Andrew Pickering, *The Mangle of Practice: Time, Agency, and Science* (Chicago: University of Chicago Press, 1995).

21. Bagemihl, *Biological Exuberance*, 471.

22. See McWhorter's brilliant recasting of deviance, which articulates sexual deviance with evolutionary deviation, resulting in a formulation that generates a queer green ethics: "It was deviation in development that produced this grove, this landscape, this living planet. What is good is that the world remain ever open to deviation." See McWhorter, *Bodies and Pleasures: Foucault and the Politics of Sexual Normalization* (Bloomington: Indiana University Press, 1999), 164.

23. Donna J. Haraway, *The Companion Species Manifesto: Dogs, People, and Significant Otherness* (Chicago: Prickly Paradigm Press, 2003).

24. Eve Kosofsky Sedgwick, *Epistemology of the Closet* (Berkeley: University of California Press, 1990), 8.

25. Stacy Alaimo, "Introduction: Feminist Theory's Flight from Nature," in

Undomesticated Ground: Recasting Nature as Feminist Space (Ithaca, N.Y.: Cornell University Press, 2000), 1–26.

26. Jonathan Marks, *What It Means to Be 98% Chimpanzee: Apes, People, and Their Genes* (Berkeley: University of California Press, 2002), 110.

27. Ibid., 111.

28. Ibid., 165.

29. Ibid.

30. Jennifer Terry, "'Unnatural Acts' in Nature: The Scientific Fascination with Queer Animals," *GLQ* 6, no. 2 (2000): 152.

31. Ibid., 154.

32. Ibid., 151.

33. Ibid., 185.

34. Haraway, *Companion Species Manifesto*, 5.

35. Haraway, *Primate Visions*, 8.

36. Ibid., 8.

37. See the essays within *Material Feminisms* for a range of approaches that combine postmodernism, poststructuralism, and social construction with a commitment to productively engaging with the materiality of human bodies and more-than-human natures and environments. Stacy Alaimo and Susan J. Hekman, "Introduction," in *Material Feminisms*, ed. Alaimo and Hekman (Bloomington: Indiana University Press, 2008), 1–22. Hekman's essay in *Material Feminisms*, "Constructing the Ballast: An Ontology for Feminism," 85–119, provides an excellent map of four different "settlements" in contemporary theory in which this new paradigm is emerging.

38. Cynthia Chris, *Watching Wildlife* (Minneapolis: University of Minnesota Press, 2006), 156.

39. Ibid., 157.

40. Ibid., 165.

41. See, for example, Evelyn Fox Keller's critique of genetic determinism in *The Century of the Gene* (Cambridge, Mass.: Harvard University Press, 2002). Another striking counterpoint to genetic determinism would be Ronnie Zoe Hawkins's contention that "the message of the genome is the opposite of biological determinism: our primate biology provides us with a tremendous amount of behavioral flexibility, while our social and cultural environments are often in the role of maintaining practices that have become maladaptive." See Hawkins, "Seeing Ourselves as Primates," *Ethics and the Environment* 7, no. 2 (2002): 60–61.

42. Roger N. Lancaster, *The Trouble with Nature: Sex in Science and Popular Culture* (Berkeley: University of California Press, 2003), xi.

43. Ibid., 29.

44. See Lynda Birke's discussion of how most critiques of biological determinism apply only to humans, which means that they not only ignore the

behavior of other animals, but rely on a strict human/animal dichotomy. Birke, *Feminism, Animals, and Science: The Naming of the Shrew* (Buckingham, U.K.: Open University Press, 1994).

45. Lancaster, *Trouble with Nature*, 61.
46. My facile division of this terminology raises larger epistemological and ethical questions regarding the discourses for animal sex. Terms that seem too anthropomorphic disrespect the "differences" of various nonhuman creatures. Terms that seem too anti-anthropomorphic shore up the human/animal divide, casting animals as mechanistic creatures of instinct or genetic determinism. There is no way out of this dilemma; our terms are strands within these webs of meaning.
47. Ibid., 31.
48. Ibid., 114.
49. One aspect of the new materialism in science studies, or of "material feminisms" (see Alaimo and Hekman, "Introduction"), is an openness to the transgressive, progressive potential for theoretical engagements with materiality. Myra J. Hird puts it succinctly in "Naturally Queer": "We may no longer be certain that it is nature that remains static and culture that evinces limitless malleability." See Hird, "Naturally Queer," *Feminist Theory 5*, no. 1 (2004): 88. Roughgarden states, "Biology need not be a purveyor of essentialism, of rigid universals. Biology need not limit our potential" (*Evolution's Rainbow*, 180). In *Undomesticated Ground* (p. 17) I discuss a range of women writers, from the late nineteenth century to the present, who challenge the conception of nature as a ground of fixed essences, rigid sexual difference, and already apparent norms, values, and prohibitions.
50. Kim TallBear, "An Indigenous Reflection on Working beyond the Human / Not Human," *GLQ* 21, nos. 2–3 (2015): 234.
51. Ibid., 235.
52. Kristin L. Field and Thomas Waite, "Absence of Female Conspecifics Induces Homosexual Behavior in Male Guppies," *Animal Behavior* 68 (2004): 1381; citing J. C. Woodson, "Including 'Learned Sexuality' in the Organization of Sexual Behavior," *Neuroscience and Biobehavioral Reviews*, January 26, 2002, 69–80.
53. *Grizzly Man*, dir. Werner Herzog (Lions Gate, 2005).
54. Gayle Rubin, "The Traffic in Women: Notes on the 'Political Economy' of Sex," in *Women, Class, and the Feminist Imagination: A Socialist–Feminist Reader*, ed. Karen V. Hansen and Ilene J. Phillipson (Philadelphia: Temple University Press, 1990), 74–113.
55. Bagemihl, *Biological Exuberance*, 45.
56. Ibid., 66.
57. Ibid., 67.
58. Ibid., 69.

59. Ibid., 71.
60. Ibid., 70–71.
61. Lancaster, *Trouble with Nature*, 266.
62. Ibid.
63. Paul L. Vasey, "Sex Differences in Sexual Partner Acquisition, Retention, and Harassment during Female Homosexual Consortships in Japanese Macaques," *American Journal of Primatology* 64, no. 4 (2004): 399.
64. Alan F. Dixon, *Primate Sexuality: Comparative Studies of the Prosimians, Monkeys, Apes, and Human Beings* (Oxford, U.K.: Oxford University Press, 1988), 147.
65. Frans de Waal writes that some "authors and scientists are so ill at ease [with the bonobo's sexuality] that they talk in riddles. . . . It's like listening to a gathering of bakers who have decided to drop the word 'bread' from their vocabulary, making for incredibly circumlocutory exchanges. The sexiness of bonobos is often downplayed by counting only copulations between adults of the opposite sex. But this really leaves out most of what is going on in their daily lives. It is a curious omission, given that the 'sex' label normally refers to any deliberate contact involving the genitals, including petting and oral stimulation." See de Waal, *Our Inner Ape: A Leading Primatologist Explains Why We Are Who We Are* (New York: Riverhead, 2005), 93.
66. Paul L. Vasey, "Pre- and Postconflict Interactions between Female Japanese Macaques during Homosexual Consortships," *International Journal of Comparative Psychology* 17 (2004): 351.
67. Meagan K. Shearer and Larry S. Katz, "Female–Female Mounting among Goats Stimulates Sexual Performance in Males," *Hormones and Behavior* 50 (2006): 36.
68. Vasey, "Sex Differences in Sexual Partner Acquisition," 399.
69. Similarly, Cynthia Chris argues that within television wildlife shows homosexuality is "not a natural act to be understood on its own terms, but a phase of foreplay prior to the real reproductive deal, an assertion of power, or an experience through which one risks subordination. Pleasure for these creatures, is strictly on the rocks." See Chris, *Watching Wildlife*, 165.
70. Catriona Sandilands, "Desiring Nature, Queering Ethics: Adventures in Erotogenic Environments," *Environmental Ethics* 23, no. 2 (2001): 176. Queer animals may disrupt the prevalent marketing of "nature" as the quintessentially wholesome (straight) family recreational site. Just as I always wonder every time I teach Whitman's "Song of Myself" what decades of schoolchildren (and their teachers) thought about that blatant homosexual moment within the poem, I wonder how dolphin-tour operators respond to the question "What are they doing?!" when say, a group of male dolphins, prominent penises very much in plain sight, rub against each other in a frenzy of pleasure, right next to the tour boat. Oh, to have access to an archive of these conversations!

71. Ibid., 180.
72. Catriona Sandilands, "Sex at the Limits," in *Discourses of the Environment*, ed. Eric Darier (Oxford, U.K.: Blackwell), 92–93.
73. Karen Barad, "Posthumanist Performativity: Toward an Understanding of How Matter Comes to Matter," *SIGNS* 28, no. 3 (2003): 803.
74. Hird, "Naturally Queer," 85.
75. Ibid., 86.
76. Ibid., 85.
77. Paul. L. Vasey et al., "Male–Female and Female–Female Mounting in Japanese Macaques: A Comparative Study of Posture and Movement," *Archives of Sexual Behavior* 35, no. 2 (2006): 127.
78. Ibid., 126.
79. Judith Butler, *Gender Trouble: Feminism and the Subversion of Identity* (New York: Routledge, 2006); Judith Halberstam, *Female Masculinity* (Durham, N.C.: Duke University Press, 1998).
80. Vasey et al., "Male–Female and Female–Female Mounting," 127.
81. Roughgarden, *Evolution's Rainbow*, 27.
82. Ibid., 28.
83. Ibid., 9.
84. Haraway, *Companion Species Manifesto*, 25.
85. Roughgarden, *Evolution's Rainbow*, 138.
86. Ibid.
87. Ibid., 37.
88. Bagemihl, *Biological Exuberance*, 260–61.
89. Ana Mariella Bacigalupo, *Shamans of the Foye Tree: Gender, Power, and Healing among Chilean Mapuche* (Austin: University of Texas Press, 2007), 52.
90. TallBear, "An Indigenous Reflection," 234–35.
91. Celia Lowe, *Wild Profusion: Biodiversity Conservation in an Indonesian Archipelago* (Princeton, N.J.: Princeton University Press, 2006), 160.
92. Ibid., 107.
93. Heather Houser, *Ecosickness in Contemporary U.S. Fiction: Environment and Affect* (New York: Columbia University Press, 2014), 78–79.
94. Jeffrey Jerome Cohen, *Stone: An Ecology of the Inhuman* (Minneapolis: University of Minnesota Press, 2015), 9.
95. This alludes to the J. B. S. Haldane quote "The Universe is not only queerer than we suppose, it is queerer than we can suppose," which Bagemihl, Hird, and Lancaster use as an epigraph.
96. I should point out here that contemplating nonhuman sexual diversity can provoke rather unproblematic modes of wonder, compared to, say, the sort of "wonder" one might feel when considering the paralyzing complexities of climate change or the unthinkable enormity of the sixth great extinction, or the epistemologically fraught environments, teeming with often unknown dangers, that we dwell within. It is important

to note that Heather Houser complicates the argument about wonder cited above, noting how Richard Power's novel *The Echo Maker* not only demonstrates the potency of wonder for environmentalism but shows how the same "defamiliarizing vacillations" that "may produce wonder" also "introduce paranoia and projection and obfuscate pathways between perception and care." Houser, *Ecosickness in Contemporary U.S. Fiction*, 115. This brief note does not encapsulate her nuanced reading.

97. Pickering, *Mangle of Practice*, 23.

98. Samantha Frost offers a lucid explanation of the sort of environmentally oriented stance of epistemological humility that I have long advocated: "What is at stake in thinking in terms of complexity, interdependence, and ecology broadly construed is epistemological and political humility in the face of the organic and inorganic world: an acknowledgement of the impossibility of full and definitive knowledge and a corollary surrender of the teleological assumption that we might possibly, at some future point, achieve full mastery over ourselves and the world around us." Such a dream becomes ironic in the anthropocene, which marks a brutally ironic sort of mastery that was, of course, no one's intended outcome. I also agree with Frost's observation that "to acknowledge a zone of necessary ignorance in complexity is not tantamount to giving up on knowledge altogether: we do not need the promise of full knowledge as the backdrop for scientific investigations." See Frost, "The Implications of the New Materialisms for Feminist Epistemology," in *Feminist Epistemology and Philosophy of Science: Power in Knowledge*, ed. H. E. Grasswick (Berlin: Springer, 2011), 79.

99. It is somewhat ironic to draw on the term "companion species," here, which Haraway has developed primarily with domesticated dogs as the model. Not only does Bagemihl's bestiary rank the homosexuality of wild dogs as "incidental," rather than "moderate or primary," but most domesticated dogs are routinely deprived of their sexuality. Karla Armbruster argues, notwithstanding the overpopulation of dogs, that "while in our quest for innocence, we have radically suppressed and made invisible much of the wildness that could be expressed by dogs through their sexuality and reproduction, much is lost. For the dogs, one more opportunity to exercise their own agency and instincts has been denied them, limiting their chances for flourishing." She notes that if companion animals are to "lead us back to a more intimate relationship with the natural world," we would "need to be open to their otherness, their wildness, the very aspects of their lives and behavior that challenge the norms of culture, that surprise us, disgust us, scare us, or (if we let them) energize us and even fill us with wonder." See Armbruster, "Into the Wild: Response, Respect, and the Human Control of Canine Sexuality and Reproduction," *JAC* 30, nos. 3–4 (2010): 763.

100. Haraway, *Companion Species Manifesto*, 6.
101. Brian Massumi, *Parables for the Virtual: Movement, Affect, Sensation* (Durham, N.C.: Duke University Press, 2002), 27.
102. Volker Sommer, "'Against Nature?!' An Epilogue about Animal Sex and the Moral Dimension," in *Homosexual Behavior in Animals: An Evolutionary Perspective*, ed. Volker Sommer and Paul L. Vasey (Cambridge, U.K.: Cambridge University Press, 2006), 370.
103. Duane Jeffrey, "Review of Joan Roughgarden's *Evolution's Rainbow*," *Politics and the Life Sciences* 23, no. 2 (2005): 72.
104. Roughgarden, *Evolution's Rainbow*, 2.
105. Bagemihl, *Biological Exuberance*, 1.
106. Bruno Latour, "Why Has Critique Run Out of Steam? From Matters of Fact to Matters of Concern," *Critical Inquiry* 30 (Winter 2004): 237.
107. Bagemihl, *Biological Exuberance*, 6.

3. The Naked Word

1. Stacy Kalish, "The Naked War," Bare Witness website, 2003, http://swaporamarama.org/images/nobushpress/Bare%20Witness%20Stacy%20Article.pdf.
2. Chris Steenberg, "Peace Protest on Icy Forest," *East Ginstead Courier*, October 2003, reprinted on the Bare Witness website, http://www.barewitness.org/news.html.
3. PETA (People for the Ethical Treatment of Animals), "Join the Human Race: Running of the Nudes: Out with the Old, in with the Nude," n.d., http://www.runningofthenudes.com/.
4. Sebastian Malo, "Naked Outdoor Protest over SeaWorld Float in NY's Thanksgiving Parade," Reuters, November 19, 2014, http://www.reuters.com/article/usa-thanksgiving-seaworld-idUSL2N0T92U120141119.
5. Samantha Ketterer, "Topless Woman Takes on Open-Carry Supporters in Austin," *Chron*, March 11, 2016, http://www.chron.com/news/houston-texas/texas/article/Austin-weird-Topless-woman-takes-on-open-carry-6884779.php.
6. Thanks to J. Halberstam for telling me about this group.
7. See Stacy Alaimo and Susan J. Hekman, eds., *Material Feminisms* (Bloomington: Indiana University Press, 2008).
8. Gloria Anzaldúa, *Borderlands / La Frontera: The New Mestiza* (San Francisco: Aunt Luke Books, 1987). In Ana Louise Keating's fascinating essay on Anzaldúa she writes, "Drawing on indigenous theories of participatory language and Gloria Anzaldúa's work, I develop a transformation-based writing practice that I call poet–shaman aesthetics: a synergistic combination of artistry, healing, and transformation grounded in relational, indigenous-inflected worldviews. I focus especially on the physi-

cal dimensions of Anzaldúa's writing, where the words she uses, the metaphors she creates, emerge from and connect with her subjects and have physiological and other material effects. Poet–shaman aesthetics represents an entirely embodied and potentially transformative intertwining of language, physiology/matter, and world." See Keating, "Speculative Realism, Visionary Pragmatism, and Poetic–Shamanic Aesthetics in Gloria Anzaldúa—and Beyond," *WSQ: Women's Studies Quarterly* 40, nos. 3–4 (2012): 51.

9. Gail Weiss, *Body Images: Embodiment as Intercorporeality* (New York: Routledge, 1999). In her definitions of the term, Weiss does not restrict intercorporeality to the human; rather, she argues that the "bodily imperative" extends to "bodies that are not human such as animal bodies, bodies of literature, and technological bodies" (163). The bulk of her analyses and arguments, however, are concerned with human intercorporeality, which makes sense, given that the formation of "body image" is her focus. I should also distinguish between how I'm using the term "transcorporeality" in the context of this chapter from how it is developed in my *Bodily Natures: Science, Environment, and the Material Self* (Bloomington: Indiana University Press, 2010). In that work it emerges from environmental health and environmental justice, and involves both a science studies conception of the "material captures" necessary in risk society and an emphasis on intra-active material agencies drawn from Karen Barad's work. The political performance of a different sort of trans-corporeality and the potential for experiencing intimacy across human bodies and physical places as an incitement to ethics is my focus here.

10. Hal Foster, *Recodings: Art, Spectacle, Cultural Politics* (Seattle: Bay Press, 1985), 89.

11. "Climate Camp Protests Target RBS and Shell in Central London," *The Telegraph*, n.d., http://www.telegraph.co.uk/news/earth/earthpicturegalleries/6122948/Climate-Camp-protests-target-RBS-and-Shell-in-Central-London.html?image=7.

12. "Naked Protest," *Wells Journal*, April 3, 2003.

13. Bare Witness, n.d., http://www.barewitness.org/.

14. For a revealing account of the interconnections between "queer primitivism" and pro-sex feminism, as well as an incisive critique of the failure of these movements to "question their desire to occupy primitivity, which in a commoditization of indigeneity and naturalization of conquest grants them authority over past and future worlds of sexual essence and authentic culture," see Scott Morgensen, "Rooting for Queers: A Politics of Primitivity," *Women and Performance* 29, no. 15 (2005): 276.

15. Earth First! Listserv, January 31, 2001.

16. *Striptease to Save the Trees*, dir. James Ficklin and K. Rudin (Earth Films, 2000). An excerpt of the film is available at https://vimeo.com/3448925.

17. Similarly, Ruth Barcan argues that female nudism opens up a "different kind of space . . . a space involving quite distinctive intercorporeal relations which help to produce new body images." See Barcan, "The Moral Bath of Bodily Unconsciousness: Female Nudism, Bodily Exposure and the Gaze," *Continuum: Journal of Media and Cultural Studies* 15, no. 3 (2001): 314.

18. Peggy Phelan, *Unmarked: The Politics of Performance* (New York: Routledge, 1993), 6.

19. Timothy Luke, "Reconstructing Nature: How the New Informatics Are Rewriting the Environment and Society as Bitspace," *Capitalism, Nature, Socialism* 47 (September): 3–28.

20. Arturo Escobar, "Gender, Place and Networks: A Political Culture of Cyberculture," in *Women@Internet: Creating New Cultures in Cyberspace*, ed. Wendy Harcourt (London: Zed, 1999), 36.

21. Ibid., 52.

22. Richard Grusin, *Premediation: Affect and Mediality after 9/11* (New York: Palgrave, 2010), 3.

23. Barcan, "Moral Bath of Bodily Unconsciousness," 314.

24. Mar del Plata, Argentina, n.d., Bare Witness website, http://www.bare witness.org/.

25. Hortense Spillers, "Mama's Baby, Papa's Maybe: An American Grammar Book," *Diacritics* 17, no. 2 (1987): 67.

26. Ibid. Commenting on this quote by Spillers, Amber Jamilla Musser writes, "As such, flesh occupies a fraught position within studies of difference. It oscillates between being a symptom of abjection and objectification and a territory ripe for reclamation." See *Sensational Flesh: Race, Power, and Masochism* (New York: NYU Press, 2014), 20.

27. Ibid.

28. Ibid., 76.

29. Nicole Fleetwood, *Troubling Vision: Performance, Visuality, and Blackness.* (Chicago: University of Chicago Press, 2011), 9, 112.

30. Ibid.

31. Ibid., 112.

32. Alexander G. Weheliye, *Racializing Assemblages: Biopolitics, and Black Feminist Theories of the Human* (Durham, N.C.: Duke University Press, 2014), 8.

33. Ibid., 32.

34. Ibid., 44–45.

35. In any case, as I revise this chapter in 2016, I am troubled by how the content, even as it pertains to serious environmental concerns, is eclipsed by recent events in the United States: the police brutality against African Americans, an intensification of institutional and other forms of racism, the exploitation of African Americans in the prison industrial complex, and long-term economic and educational injustice.

36. Helena, Montana, January 25, 2003, Bare Witness website, http://www
.baringwitness.org/Helena.htm.

37. Rebecca Schneider, *The Explicit Body in Performance* (New York: Routledge,
1997), 67.

38. Diana Saco, *Cybering Democracy: Public Space and the Internet* (Minneapolis:
University of Minnesota Press, 2002), 27.

39. Phelan, *Unmarked*, 69.

40. See Stacy Alaimo, *Undomesticated Ground: Recasting Nature as Feminist
Space* (Ithaca, N.Y.: Cornell University Press, 2000).

41. Luce Irigaray, *This Sex Which Is Not One*, trans. Catherine Porter (Ithaca,
N.Y.: Cornell University Press, 1985), 77.

42. John Berger, *Ways of Seeing* (London: British Broadcasting Corporation,
1985), 47.

43. Donna Haraway, "The Promise of Monsters: A Regenerated Politics for
Inappropriate/d Others," in *Cultural Studies*, ed. Lawrence Grossberg,
Cary Nelson, and Paula Treichler (New York: Routledge, 1992), 313.

44. Plumwood explains: "One of the most common forms of denial of
women and nature is what I will term backgrounding, their treatment
as providing the background to a dominant, foreground sphere of recognized achievement or causation. This backgrounding of women and
nature is deeply embedded in the rationality of the economic system and
in the structures of contemporary society." See *Feminism and the Mastery
of Nature* (New York: Routledge, 2003), 21.

45. Phelan, *Unmarked*, 146.

46. Elizabeth Bray and Claire Colebrook, "The Haunted Flesh: Corporeal Feminism and the Politics of (Dis)embodiment," *SIGNS* 24, no. 1 (1998): 44.

47. Catriona Sandilands, *The Good-Natured Feminist: Ecofeminism and the Quest
for Democracy* (Minneapolis: University of Minnesota Press, 1999), 180.

48. Ibid.

49. As a counterpoint to my reading which extends Phelan's arguments, see
Meileng Cheng's discussion of the Sacred Naked Nature Girls performance group, in which she argues that "presence (defined as representa-
tional visibility and audibility) still offers more possibility than absence,"
contending that "we must reclaim the corporeal attributes of presence."
Cheng, *In Other Los Angeles: Multicentric Performance Art* (Berkeley: University of California Press, 2002), 246.

50. Weiss, *Body Images*, 5.

51. *Striptease to Save the Trees*, dir. Ficklin and Rudin.

52. I develop the term "trans-corporeality" more fully in *Bodily Natures*.

53. Moira Gatens, *Imaginary Bodies: Ethics, Power, and Corporeality* (New York:
Routledge, 1996), 110.

54. "Blondes Go Naked to Save Animals," *Gold Coast Bulletin*, March 2, 2002.

55. "Bare Witness News," n.d., Bare Witness website, http://www.barewitness .org/.

56. "World Naked Bike Ride: Hordes of Nude Cyclists Set to Descend on London for Protest against Car Culture," *Evening Standard*, June 12, 2015, http://www.standard.co.uk/news/london/world-naked-bike-ride -hordes-of-nude-cyclists-set-to-descend-on-london-for-protest-against -car-10314959.html.

57. "!!ACTIVIST GET NUDE FOR CLIMATE CHANGE!!" YouTube, October 4, 2009, https://www.youtube.com/watch?v=9EN16stS6nk.

58. "600 Strip Naked on Glacier in Global Warming Protest," August 8, 2007, Greenpeace website, http://www.greenpeace.org/international/en/ news/features/naked-glacier-tunick-08182007/.

59. Ibid.

60. Ibid.

61. "Baring Witness," n.d., Bayside Productions, http://www.baringwitness .org/.

62. Quoted in Kalish, "Naked War."

63. Ibid.

64. Bonnie Mann, "How America Justifies Its War: A Modern/Postmodern Aesthetics of Masculinity and Sovereignty," *Hypatia* 21, no. 4 (2006): 155.

65. Ibid., 159.

66. Anita Roddick, "Dispatch: Stripping for Peace," June 2003, http://www .anitaroddick.com/.

67. "Peace," January 2003, Bare Witness website, http://www.barewitness.org/.

68. Stuart Jeffries, "Anti-war Strippers Brave the Ice," *The Guardian*, January 14, 2003, http://www.theguardian.com/theguardian/2003/jan/14/ features11.g2.

69. John Barry, "Vulnerability and Virtue: Democracy, Dependency, and Ecological Stewardship," in *Democracy and the Claims of Nature: Critical Perspectives for a New Century*, ed. Ben A. Minteer and Bob Pepperman Taylor (Lanham, Md.: Rowman & Littlefield, 2002), 133.

70. The passage, which is more complex than what is represented in the body of the chapter, deserves quoting at length: "And if we understand ourselves as *weathering*, intra-actively made and unmade by the chill of a too-cold winter, the discomfort of a too-hot sun, then we can also attune ourselves to the pasts that are contracted in changing temperatures, rising sea levels, increasingly desiccated earths. We attune ourselves to the singularities of its intra-actions, recognizing the multitude of bodies (including our own) that are all co-emerging in the making of these weather-times. We recognize our own implications in the climatic conditions around us, thick with co-labored temporalites that we are also making possible." Astrida Neimanis and Rachel Loewen Walker, "Weathering: Climate Change and the 'Thick Time' of Transcorporeality," *Hypatia* 29, no. 3 (2014): 559, 573.

71. Fuck for Forests, n.d., http://www.fuckforforest.com/en/ecoarbofilia.html.

72. Gregory Dicum, "Green Ecoporn: Great Sex for a Good Cause," *SF Gate*, April 13, 2005, http://www.sfgate.com/homeandgarden/article/GREEN-Eco-porn-Great-Sex-For-A-Good-Cause-3175838.php.

73. Fuck for Forest website, n.d., http://www.fuckforforest.com/en/about.html.

74. *Striptease to Save the Trees*, dir. Ficklin and Rudin.

75. See Alaimo, "Emma Goldman's Mother Earth and the Nature of the Left," in *Undomesticated Ground*.

76. Emma Goldman and Max Baginski, "Mother Earth," *Mother Earth* 1, no. 1 (1906): 2.

77. Irigaray, *This Sex Which Is Not One*, 133.

78. Catriona Sandilands, "Sex at the Limits," in *Discourses of the Environment*, ed. Eric Darier (Oxford, U.K.: Blackwell, 1999), 77.

79. Ibid., 78.

80. Jeanie Forte, "Women's Performance Art: Feminism and Postmodernism," in *Performing Feminisms: Feminist Critical Theory and Theatre*, ed. Sue-Ellen Case (Baltimore: Johns Hopkins University Press, 1990), 262.

81. Environmental justice movements recast environmental politics to include places where people work, and analyze the distribution of environmental harms and benefits by way of race and class. For more on the issue of environmentalism and labor see Richard White, who asks, "How is it that environmentalism seems opposed to work? And how is it that work has come to play such a small role in American environmentalism?" White, " 'Are You an Environmentalist or Do You Work for a Living?': Work and Nature," in *Uncommon Ground: Rethinking the Human Place in Nature*, ed. William Cronon (New York: Norton, 1996), 171. See also Catriona Sandilands, "Between the Local and the Global," in which she argues, "Forest workers, despite their frequently long-term and intimate interactions with the forest ecosystem, do not count as knowing nature because the only real knowledge of nature is a consumptive one." Sandilands, "Between the Local and the Global: Clayoquot Sound and Simulacral Politics," in *A Political Space: Reading the Global through Clayoquot Sound*, ed. Warren Magnuson and Karena Shaw (Minneapolis: University of Minnesota Press. 2003), 157.

82. Phelan, *Unmarked*, 165.

83. Urban Dictionary, "ecosexual," n.d., http://www.urbandictionary.com/define.php?term=ecosexual; Katy Neusteter, "Quiz: Are You an Eco-Sexual?" *Gaiam Life*, n.d., http://life.gaiam.com/article/quiz-are-you-eco-sexual.

84. Beth Stephens and Annie Sprinkle, "Ecosex Manifesto," n.d., Sexecology website, http://www.sexecology.org/.

85. "Here Come the Ecosexuals," n.d., http://theecosexuals.ucsc.edu/.

86. Linda S. Kaufman, *Bad Girls and Sick Boys: Fantasies in Contemporary Art and Culture* (Berkeley: University of California Press, 1998), 58.
87. Ibid., 58–59.
88. Sexecology website, n.d., http://www.sexecology.org/.
89. Tim Ferguson, quoted on the World Wide Nudism & Naturism News site, n.d., http://www.worldwidenudismnaturism.com/pages/news/protest.html.
90. "Buddha Bear," posting on the Backpacker website, n.d., http://www.backpacker.com/.
91. Jane Kay, "Topless Protestor Bemuses, Confuses," *Examiner*, http://www.tree-sit.org/tiger.html.
92. Patricia Miller, "The Rhetorical Web," *Anvil*, March 2001, http://www.anvilmedia.com/archives/.
93. Naomi Jarvie, "'La Tigresa, Nude Savior of the Forest' Holds FB Rally," *Fort Bragg Advocate News*, 2001, reprinted on the Mendocino Redwood Company's website, http://www.mrc.com/News/news021501.htm.
94. Nieto, quoted in "Asides," *Pittsburgh Post-Gazette*, October 22, 2000.
95. "Naked Protest against War," *Manila Bulletin*, March 6, 2003.
96. Wendy Parkins, "Protesting Like a Girl: Embodiment, Dissent, and Feminist Agency," *Feminist Theory* 1, no. 1 (2000): 73.
97. Emily Martin, "Fluid Bodies, Managed Nature," in *Remaking Reality: Nature at the Millennium*, ed. Bruce Braun and Noel Castree (New York: Routledge, 1988), 79.
98. Ibid., 78.
99. Weiss, *Body Images*, 141.
100. Judith Stacey, "The Empress of Feminist Theory Is Overdressed," *Feminist Theory* 2, no. 1 (2001): 102.

4. Climate Systems, Carbon-Heavy Masculinity, and Feminist Exposure

1. Kirsten Justesen, Ice Pedestal Formations #1, 2000, http://www.kirstenjustesen.com/index.php?id=332. For more essays from the Gendering Climate Change and Sustainability conference at the University of Denmark, 2009, see *Women, Gender and Research (Kvinder, Kon og Forskning)* 2–3 (2009): 22–35. This issue was distributed at the COP 15 international climate change summit in December 2009. I was honored to be invited to speak at the Gender and Climate Change conference in Copenhagen and especially pleased with the way the conference brought together feminist theorists and other academics with policy makers, government officials, and NGO officials. The women who drafted some of the feminist climate change documents that I critiqued heard my talk and received the criticisms not only with grace but with the determination to construct posi-

tions that would work for feminist and queer politics and, at the same time, have cultural purchase in broader cultural, political, and governmental contexts. I very much admire and respect the political work that they do.

2. Esther Adler, "Kirsten Justesen: My Body as Material: Danish Artist Kirsten Justesen Speaks with MoMA Curatorial Assistant Esther Adler," *P.S.1 Newspaper*, 2008, http://momaps1.org/images/pdf/newspaper/08spring/Newspaper_Winter%202008_OK.pdf.

3. Ibid.

4. The three Ice Plinth photographs can be seen on Kirsten Justesen's website at http://www.kunst-paa-arbejdspladsen.dk/category/kunstnere/kirsten-justesen.

5. And of course there are many feminist protests that are not about environmental issues, but which strategically employ nudity. For example, the feminist group that began in Ukraine, FEMEN, whose Facebook description is "Our God is a Woman! Our Mission is Protest! Our Weapon are bare breasts," protests for women's sexual, reproductive, and religious freedom. See Facebook, FEMEN International, https://www.facebook.com/FEMEN.International.Official/info?tab=page_info; and FEMEN's website, http://femen.org/news.

6. "600 Strip Naked on Glacier in Global Warming Protest," August 8, 2007, Greenpeace website, http://www.greenpeace.org/international/en/news/features/naked-glacier-tunick-08182007/.

7. There are several important essays on this topic, including Giovanna Di Chiro, "Polluted Politics? Confronting Toxic Discourse, Sex Panic, and Eco-Normativity," in *Queer Ecologies: Sex, Nature, Politics, Desire*, ed. Catriona Mortimer-Sandilands and Bruce Erickson (Bloomington: Indiana University Press, 2010), 331–58.

8. The politics of exposure parallels my concept of trans-corporeality, although trans-corporeality also involves a sense of intra-active material agencies, which I do not discuss in this chapter.

9. Bonnie Mann, "How America Justifies Its War: A Modern/Postmodern Aesthetics of Masculinity and Sovereignty," *Hypatia* 21, no. 4 (2006): 147–63.

10. Susan Jeffords, *Hard Bodies: Hollywood Masculinity in the Reagan Era* (New Brunswick, N.J.: Rutgers University Press, 1993).

11. Ibid., 25.

12. Margot Adler, "Behind the Ever-Expanding American Dream House," *All Things Considered*, July 4, 2006, http://www.npr.org/templates/story/story.php?storyId=5525283&from=mobile.

13. Ibid.

14. "Rolling coal on hot babe," YouTube, November 18, 2011, https://www.youtube.com/watch?v=iRu3xtTH9ng.

15. David Weigel, "Rolling Coal," *Slate*, July 3, 2014, http://www.slate.com/

articles/news_and_politics/politics/2014/07/rolling_coal_conservatives
_who_show_their_annoyance_with_liberals_obama.html.

16. Merritt Polk, "Gendering Climate Change through the Transport Sector," *Women, Gender and Research (Kvinder, Kon og Forskning*, Denmark) 2–3 (2009): 77.

17. Mann, "How America Justifies Its War," 148.

18. Ibid., 149.

19. Ibid., 159.

20. See Ernesto Laclau and Chantal Mouffe, *Hegemony and Socialist Strategy* (London: Verso, 1985) and many writings by Stuart Hall.

21. Judith Halberstam, *Female Masculinity* (Durham, N.C.: Duke University Press, 1988), 272.

22. Jasbir Puar, *Terrorist Assemblages: Homonationalism in Queer Times* (Durham, N.C.: Duke University Press, 2007), 40.

23. Ibid., 2.

24. Donna J. Haraway, "Situated Knowledges: The Science Question in Feminism and the Privilege of Partial Perspective," in *Simians, Cyborgs and Women: The Reinvention of Nature* (New York: Routledge, 1991), 191.

25. This section and the next focus on the time period of the George W. Bush administration, 2001–2009, since the essay was originally written as a plenary address for the Gendering Climate and Sustainability conference sponsored by the Coordination for Gender Research, University of Copenhagen, Denmark, March 13–14, 2009. While the U.S. Environmental Protection Agency has made many changes under Barack Obama's presidency, and while the many other international governmental and nongovernmental agencies have made many changes since then, I trust that the broader thrust of the critiques within this chapter will remain useful as it pertains to entrenched positions and epistemologies that are unlikely to suddenly be swept away.

26. U.S. EPA website, "Health and Environmental Effects," 2009, in "Climate Change," http://www.epa.gov/climate-change/effects/index.html.

27. World Health Organization, "Climate Change and Human Health," 2009, http://www.who.int/hac/techguidance/preparedness/areas_of_work/en/.

28. U.S. EPA, "Ecosystems and Biodiversity," 2009, in "Climate Change," http://www.epa.gov/climatechange/effects/eco.html.

29. Robert Proctor, *Cancer Wars: How Politics Shapes What We Know and Don't Know about Cancer* (New York: Basic, 1996), 8.

30. U.S. EPA website, "Global Earth Observation System of Systems," 2009, http://www.epa.gov/geoss/fact_sheets/earthobservation.html.

31. Spencer R. Weart, *The Discovery of Global Warming* (Cambridge, Mass.: Harvard University Press, 2008), ix.

32. U.S. EPA, "Global Earth Observation System of Systems."

33. Ibid.

34. Mette Bryld and Nina Lykke, *Cosmodolphins: Feminist Cultural Studies of Technology, Animals and the Sacred* (London: Zed, 2000), 6.
35. Group on Earth Observations, "What Is GEOSS: The Global Earth Observation System of Systems," 2009, http://www.earthobscrvations.org/geoss.html.
36. Bruno Latour, *Pandora's Hope: Essays on the Reality of Science Studies* (Cambridge, Mass.: Harvard University Press, 1999).
37. Sandra Harding, *Science and Social Inequality: Feminist and Postcolonial Issues* (Urbana: University of Illinois Press, 2006), 141.
38. Seema Arora-Jonsson, in an essay appearing after the earlier version of this chapter was published, provides a compelling, detailed critique of the assumption that women are more vulnerable to climate change and more environmentally virtuous. See "Virtue and Vulnerability: Discourses on Women and Climate Change," *Global Environmental Change* 21, no. 2 (2011): 744–51.
39. Women's Environment and Development Organization, "Climate Change," 2004, http://www.wedo.org/category/learn/campaigns/climatechange.
40. Fifty-second Session on the Commission of the Status of Women, *Gender Perspectives on Climate Change: Issues Paper* (New York: United Nations, 2008), http://www.un.org/womenwatchdaw/csw/csw52/issuespapers/Gender%2and%climate%20change%20paper%20final.pdf.
41. Ibid.
42. Halberstam, *Female Masculinity*, 268.
43. Fifty-second Session on the Commission of the Status of Women, *Gender Perspectives on Climate Change*. Other work on gender and climate change in the global south also poses women's vulnerability against women's knowledges. Analyzing women in Ghana, Trish Glazerbrook concludes, "Women in developing countries are especially and immediately threatened by climate change impacts, but also have knowledge to contribute to adaptation efforts if only their voices can be heard." See Glazerbrook, "Women and Climate Change: A Case Study from Northeast Ghana," *Hypatia* 26, no. 4 (2011): 777–78.
44. Kyle Powys Whyte, for example, writes of how Anishinaabe women and water have responsibilities to each other. Water is not merely a thing to be used, but an integral part of cultural and political relations: "Climate change impacts that degrade water in different ways will affect some of the core dimensions of Anishinaabe women's identities and their contribution within their communities, and will make their responsibilities to water more time-consuming and harder (if not impossible) to carry out." See Whyte, "Indigenous Women, Climate Change Impacts, and Collective Action," *Hypatia* 29, no. 3 (2014): 606.
45. Fifty-second Session on the Commission of the Status of Women, *Gender Perspectives on Climate Change*.
46. Christine Bauhardt, "Rethinking Gender and Nature from a Material(ist)

Perspective: Feminist Economics, Queer Ecologies, and Resource Politics," *European Journal of Women's Studies* 20, no. 4 (2013): 370.

47. Nina Lykke, "Non-innocent Intersections of Feminism and Environmentalism," *Women, Gender and Research (Kvinder, Kon og Forskning,* Denmark) 2–3 (2009): 36–43.

48. Hilda Rømer Christensen, Michala Hvidt Breengaard, and Helene Oldrup, "Gendering Climate Change," *Women, Gender and Research (Kvinder, Kon og Forskning,* Denmark) 2–3 (2009): 7.

49. Carolyn Hannan, "Gender Mainstreaming Climate Change," *Women, Gender and Research (Kvinder, Kon og Forskning,* Denmark) 2–3 (2009): 48.

50. Fifty-second Session on the Commission of the Status of Women, *Gender Perspectives on Climate Change.*

51. Global Gender and Climate Change Alliance, "Launch of the Global Gender and Climate Change Alliance," 2009, http://www.wedo.org/wp-content/uploads/global-gender-and-climate-alliance.pdf.

52. United Nations, "Mainstreaming Gender into the Climate Change Regime," December 14, 2004, COP 10, Buenos Aries, http://www.genanet.de/fileadmin/downloads/.

53. Halberstam, *Female Masculinity,* 27.

54. Canadian International Development Agency, "Gender Equality and Climate Change," 2009, http://www.acdicida.gc.ca/INET/IMAGES.NSF.

5. Oceanic Origins, Plastic Activism, and New Materialism at Sea

1. Callum Roberts, *The Unnatural History of the Sea* (Washington, D.C.: Island Press 2007), xv.

2. Sylvia Earle, *The World Is Blue: How Our Fate and the Oceans Are One* (Washington, D.C.: National Geographic Society, 2009), 12.

3. *Sharkwater,* dir. Rob Stewart (Sharkwater Productions and Diatribe Pictures, 2007).

4. See Earle, *World Is Blue,* for more on ocean ecologies.

5. Jonathan Safran Foer, *Eating Animals* (New York: Little, Brown, 2009), 50.

6. Stacy Alaimo, *Bodily Natures: Science, Environment, and the Material Self* (Bloomington: Indiana University Press, 2010).

7. See especially Butler's "Contingent Foundations": "'The 'I' is the transfer point of that replay, but it is simply not a strong enough claim to say that the 'I' is situated; the 'I,' this 'I,' is *constituted* by these positions, and these 'positions' are not merely theoretical products, but fully embedded organizing principles of material practices and institutional arrangements, those matrices of power and discourse that produce me as a viable 'subject.'" Butler, "Contingent Foundations: Feminism and the Question of 'Postmodernism,'" in *Feminists Theorize the Political,* ed. Judith Butler and Joan W. Scott (New York: Routledge, 1992), 9.

8. Karen Barad, *Meeting the Universe Halfway* (Durham, N.C.: Duke University Press, 2007); Ulrich Beck, *Risk Society: Towards a New Modernity*, trans. Mark Ritter (London: Sage, 1992).

9. For a new materialist approach to ecocriticism, specifically, see Serpil Oppermann's compelling essay "Ecocriticism's Theoretical Discontents," which traces the hostility toward postmodern and poststructuralist theory in ecocriticism. She argues that "we need to advance a critical perspective in which both discursivity and materiality (in other words, discursive practices and material phenomena) can be integrated in a relational approach. The accountability of such an approach must, however, lie in a correct identification of the ethical, epistemological, and ontological concerns of ecocriticism's wider interest in human and non-human systems." Oppermann, "Ecocriticism's Theoretical Discontents," *Mosaic* 44, no. 2 (2011): 155. See also Serenella Iovino and Serpil Oppermann, "Introduction: Stories Come to Matter," in *Material Ecocriticism*, ed. Iovino and Oppermann (Bloomington: Indiana University Press, 2014), 1–17.

10. The vastness of the seas has long buoyed the cultural conception of the ocean as impervious to human harm. Kimberley C. Patton, in *The Sea Can Wash Away All Evils: Modern Marine Pollution and the Ancient Cathartic Ocean* (New York: Columbia University Press, 2006), discusses how "many cultures have revered the sea, and at the same time they have made it to bear and to wash away whatever was construed as dangerous, dirty, or morally contaminating" (xi). Regardless of whether these religious beliefs have persisted, both the scale and the hazardous nature of what is dumped into the seas have changed, entirely, from ancient times. Nonetheless, contemporary global practices of dumping garbage, sewage, weapons, toxic chemicals, and radioactive waste assume that dispersing the substances or forces across the breadth and depth of the seas will make them disappear. For more on the problematic notion of "dispersing," see my essay "Dispersing Disaster: The Deepwater Horizon, Ocean Conservation, and the Immateriality of Aliens," in *Disasters, Environmentalism, and Knowledge*, ed. Sylvia Mayer and Christof Mauch (Heidelberg, Ger.: Universitätsverlag, 2012), 175–92.

11. Benthic zones are on the sea floor, while pelagic zones are in the open seas; other ocean zones are distinguished by depth.

12. See Alaimo, "Dispersing Disaster."

13. I plan to discuss this in my next book, *Blue Ecologies: Science, Aesthetics, and the Creatures of the Abyss*. For a preview, see "Violet-Black: Ecologies of the Abyssal Zone," in *Prismatic Ecologies: Ecotheory beyond Green*, ed. Jeffrey Jerome Cohen (Minneapolis: University of Minnesota Press, 2013), 233–51.

14. William Faulkner, *As I Lay Dying* (New York: Vintage, 1991), 84.

15. I am indebted to Joni Adamson and Sally L. Kitch for this section of the chapter, as they invited me to speak on "origins" for the Institute for

Humanities Research at Arizona State University. I would also like to thank Tema Genus at the Linköping University for inviting me to speak on trans-corporeality at sea.

16. Gilles Deleuze and Félix Guattari, *A Thousand Plateaus*, trans. Brian Massumi (Minneapolis: University of Minnesota Press, 1987).

17. Donna J. Haraway, *The Companion Species Manifesto: Dogs, People, and Significant Otherness* (Chicago: Prickly Paradigm Press, 2003).

18. Donna J. Haraway, *When Species Meet* (Minneapolis: University of Minnesota Press, 2008), 25.

19. Barad, *Meeting the Universe Halfway*, 139.

20. Ibid., 396.

21. Ibid.

22. Darwin, quoted in Carl Zimmer, *At the Water's Edge: Fish with Fingers, Whales with Legs, and How Life Came Ashore but Then Went Back to Sea* (New York: Simon & Schuster, 1998), 23.

23. Vilém Flusser and Louis Bec propose that "disgust recapitulates phylogenesis," as the "more disgusting something is, the further removed it is from humans on the phylogenetic tree." See Flusser and Bec, *Vampyroteuthis Infernalis: A Treatise, with a Report by the Institut Scientifique de Recherge Paranaturaliste*, trans. Valentine A. Pakis (Minneapolis: University of Minnesota Press, 2012), 11. While popular hierarchical and teleological conceptions of evolution that conclude triumphantly with humans deserve denunciation, Flusser and Bec's suggestion that Darwin should be blamed for "anthropocentric" "biological criteria," which justify human disgust by systematizing "this rationalization of the irrational," seems a willful misreading of Darwin's actual writings, which are filled with regard, affection, sympathy, curiosity, and wonder for other creatures. Moreover, Darwin needed to curtail creaturely disgust in order for his theory of human evolution to become accepted. Yet one hesitates to venture any sort of comment on *Vampyroteuthis Infernalis*, as its parodic and satiric mode leave the reader unmoored (ibid., 12).

24. Charles Darwin, *The Origin of Species by Means of Natural Selection / The Descent of Man and Selection in Relation to Sex* (New York: Modern Library, n.d.), 395–96.

25. Ibid., 411.

26. Ibid., 911.

27. In *Undomesticated Ground: Recasting Nature as Feminist Space* (Ithaca, N.Y.: Cornell University Press, 2000) I discuss how Darwinian feminist philosophers in the late nineteenth and early twentieth century take the fact that human ancestors were hermaphroditic and use it to argue for a gender-minimizing, even queer, feminism that emerges from nature. Directly countering the late twentieth-century idea that gender essentialisms were rooted in nature, they saw pervasive evolutionary transformation as evi-

dence that rigid gender oppositions were "unnatural." To what extent Darwin's work and evolutionary theory more generally can be read in terms of the development of contemporary feminisms and queer ecologies remains an open question. See also Stacy Alaimo, "Sexual Matters: Darwinian Feminism and the Nonhuman Turn," *J19: Journal of Nineteenth-Century Americanists* 1, no. 2 (2013): 390–96.

28. Linda Hogan, "Crossings," in *The Book of Medicines* (Minneapolis: Coffee House Books, 1993), 28–29.

29. Zimmer, *At the Water's Edge*, 6.

30. Ibid., 8.

31. Rachel Carson, *The Sea around Us* (Oxford, U.K.: Oxford University Press, 1950), 3.

32. Ibid., 8.

33. Ibid.

34. Ibid.

35. Ibid., 13–14.

36. Neil Shubin, *Your Inner Fish: A Journey into the 3.5-Billion-Year History of the Human Body* (New York: Vintage, 2008), 3.

37. Ibid., 41.

38. Ibid., 27.

39. Ibid., 18.

40. Darwin, *Origin of Species*, 374.

41. Shubin, *Your Inner Fish*, 113. A compelling counter to Shubin's anthropocentrism would be Eva Hayward's essay "More Lessons from a Starfish: Prefixal Flesh and Transpeciated Selves," which writes transsexuality as a "mutuality," a "shared ontology," with the starfish, particularly as they both regenerate "as an act of healing." See Hayward, "More Lessons from a Starfish: Prefixal Flesh and Transpeciated Selves," *WSQ: Women's Studies Quarterly* 36, nos. 3–4 (2008): 81. The conclusion to this book will include a brief discussion of Hayward's essay.

42. Shubin, *Your Inner Fish*, 115.

43. "Do the facts of our ancient history mean that humans are not special or unique among living creatures? Of course not. In fact, knowing something about the deep origins of humanity only adds to the remarkable fact of our existence: all of our extraordinary capabilities arose from basic components that evolved in ancient fish and other creatures. From common parts came a very unique construction. We are not separate from the rest of the living world; we are part of it down to our bones and, as we will see shortly, even our genes." Ibid., 43.

44. Ibid., 198.

45. Ibid., 201.

46. At this point there is a rich and varied literature in feminist theory, disability studies, environmental studies, queer theory, and other fields that

traces how bodily agencies interact with other material forces and cultural systems, most strikingly within material feminisms and other new materialisms. My book *Bodily Natures*, for example, poses people with multiple chemical sensitivity as the quintessential trans-corporeal subject who must contend with the deviant material agencies that her or his body registers.

47. Nick Hayes, *The Rime of the Modern Mariner* (New York: Viking, 2011).

48. Sylvia Earle, *Sea Change: A Message of the Oceans* (New York: Fawcett Columbine, 1996), 15.

49. Julia Whitty, "The Fate of the Ocean," *Mother Jones*, March–April 2006, 1.

50. Stephen Helmreich, "Human Nature at Sea," *AnthroNow* 2, no. 3 (2010): 50.

51. Mark McMenamin and Dianna McMenamin, *Hypersea: Life on Land* (New York: Columbia University Press, 1994), 23.

52. Ibid., 24.

53. Ibid., 24–25.

54. Ibid., 6, 5.

55. Ibid., 25.

56. Ibid., 3.

57. For more on newly discovered species of marine life see the Census of Marine Life global research network, at http://www.coml.org/.

58. McMenamin and McMenamin, *Hypersea*, 242.

59. Ibid., 255.

60. Carl Zimmer, "Hypersea Invasion," *Discover*, October 1, 1995, http://discovermagazine.com/1995/oct/hyperseainvasion571/?searchterm=mcmenamin.

61. The "Wingspread Statement on the Precautionary Principle," put forth in 1998, states, "While we realize that human activities may involve hazards, people must proceed more carefully than has been the case in recent history. Corporations, government entities, organizations, communities, scientists and other individuals must adopt a precautionary approach to all human endeavors. Therefore, it is necessary to implement the Precautionary Principle: When an activity raises threats of harm to human health or the environment, precautionary measures should be taken even if some cause and effect relationships are not fully established scientifically." Posted on the Science and Environmental Health Network website, http://www.sehn.org/wing.html.

62. Stefan Helmreich, *Alien Oceans: Anthropological Voyages in Microbial Seas* (Berkeley: University of California Press, 2009), 17.

63. Ibid.

64. Ibid., 52.

65. Ibid., 284.

66. Ibid.

67. Ibid., 60.

68. Ibid., 53.

69. See Stacy Alaimo, "Feminist Science Studies: Aesthetics and Entangle-

ment in the Deep Sea," in *Oxford Handbook of Ecocriticism*, ed. Greg Garrard (New York: Oxford University Press, 2014), 188–204.

70. Ibid. The reference here is to Bruno Latour, *On the Modern Cult of the Factish Gods* (Durham, N.C.: Duke University Press, 2010), 61.

71. Steve Mentz, *At the Bottom of Shakespeare's Ocean* (London: Continuum, 2009), 96.

72. The recognition that certain fishing practices were particularly destructive is hardly recent. Callum Roberts dates trawling—as well as criticism and debate about the practice—back to England in 1376. Someone complained to the king that the use of the "wondyrechaun" resulted in "great damage of the commons of the realm and the destruction of the fisheries" (*Unnatural History of the Sea*, 131–32).

73. Mentz, *Shakespeare's Ocean*, 98.

74. Barad, *Meeting the Universe Halfway*, 396.

75. Carson, *Sea around Us*, xiii.

76. Ibid., 15.

77. "How Radioactive Is the Ocean?" n.d., Woods Hole Oceanographic Institute, Center for Marine and Environmental Radiation, http://www.ourradioactiveocean.org/.

78. Ibid.

79. "A Shared Fate," dir. Hardy Jones (BlueVoice.org, 2008), http://www.bluevoice.org/webfilms_sharedfate.php.

80. Ibid.

81. See the Great Pacific Garbage Patch website, http://www.greatgarbagepatch.org/.

82. Chris Jordan, "Midway: Message from the Gyre," 2009, http://chrisjordan.com/gallery/midway/#CF000313%2018x24.

83. Jane Bennett, *Vibrant Matter: A Political Ecology of Things* (Durham, N.C.: Duke University Press, 2010), ix.

84. Oppermann, "Ecocriticism's Theoretical Discontents," 155.

85. See Alaimo, *Bodily Natures*, for more on how environmental justice and environmental health practitioners conceive of the agency of substances and objects.

86. E. L. Venrick et al., "Man-Made Objects on the Surface of the Central North Pacific Ocean," *Nature*, January 26, 1973, 271.

87. Greenpeace International, "The Trash Vortex," n.d., http://www.greenpeace.org/international/en/campaigns/oceans/pollution/trash-vortex/.

88. Jocelyn Kaiser, "The Dirt on Ocean Garbage Patches," *Science*, June 18, 2010, 1506.

89. Gay Hawkins, "Plastic Materialities," in *Political Matters: Technoscience, Democracy, and Public Life*, ed. Bruce Braun and Sarah J. Whatmore (Minneapolis: University of Minnesota Press, 2010), 119.

90. Ibid.

91. Ibid., 12.

92. Ibid., 137.
93. Ibid., 126.
94. Barad, *Meeting the Universe Halfway*, 393.
95. Ibid., 140.
96. In other words, I think Barad's agential realism and my trans-corporeality both diverge—in ways that are significant for environmentalism—from thing theory, thing power, and object oriented ontologies.
97. Jeffrey Jerome Cohen, "Introduction," in *Prismatic Ecologies*, ed. Cohen, xxiv.
98. See the Plastic Pollution Coalition website, http://plasticpollutioncoalition.org/.
99. Jonas Bennaroch, "The Ballad of the Plastic Bag," YouTube, May 30, 2012, http://www.youtube.com/watch?v=vQdpccDNB_A&list=UUBpIShXI_KHhwUdup_yHoVg&index=5&feature=plcp.
100. Rosi Braidotti, *Transpositions* (Malden, Mass.: Polity, 2006), 278.
101. Katrin Peters, "Plastic Seduction," YouTube, April 6, 2012, http://www.youtube.com/watch?v=G4JvMwem7mc&list=UUBpIShXI_KHhwUdup_yHoVg&index=7&feature=plcp.
102. Ibid.
103. Ibid.
104. Ian Bogost, *Alien Phenomenology; or, What It's Like to Be a Thing* (Minneapolis: University of Minnesota Press, 2012), 10, 30.
105. Nancy Tuana, "Viscous Porosity," in *Material Feminisms*, ed. Stacy Alaimo and Susan J. Hekman (Bloomington: Indiana University Press, 2010), 202.
106. Charles Moore, with Cassandra Phillips, *Plastic Ocean: How a Sea Captain's Chance Discovery Launched a Determined Quest to Save the Oceans* (New York: Penguin, 2011), 73.
107. Ibid., 51.
108. Mick Smith, *Against Ecological Sovereignty: Ethics, Biopolitics, and Saving the Natural World* (Minneapolis: University of Minnesota Press, 2011), 97.
109. Moore, *Plastic Ocean*, 66.
110. Ibid., 84.
111. Ibid., 120.
112. Ibid., 176.
113. Ibid., 85.
114. Ibid., 228–29.
115. Ibid., 230.
116. Ibid., 236.
117. Ibid., 253.
118. Serenella Iovino, "Naples 2008, or the Waste Land: Trash, Citizenship, and an Ethic of Narration," *Neohelicon*, September 29, 2009, 6.
119. Pam Longobardi, "Crime of Willful Neglect (for BP)," 2014, http://driftersproject.net/blog/2012/07/06/drifters-project-works/crimeofwillfulneglect_bp_sm-2/.

120. Stephanie LeMenager, *Living Oil: Petroleum Culture in the American Century* (New York: Oxford University Press, 2014), 69.
121. Longobardi, "Crime of Willful Neglect (for BP)."
122. Marina Zurkow, "Slurb," 6:00 excerpt from 17:42 (loop), color, animation and stereo sound. Music by Lem Jay Ignacio. Additional animation by Jen Kelly. Streaming Museum, http://streamingmuseum.org/slurb-2009-by -marina-zurkow/.
123. "About" section of ibid.

6. Your Shell on Acid

1. Richard Doyle, *Darwin's Pharmacy: Sex, Plants, and the Evolution of the Noösphere* (Seattle: University of Washington Press, 2011).
2. U.S. EPA, "Global Earth Observation System of Systems," 2009, http:// www.epa.gov/geoss/fact_sheets/earthobservation.html. These images seem to have been taken down.
3. Will Stefan, Paul J. Crutzen, and John R. McNeill, "The Anthropocene: Are Humans Now Overwhelming the Great Forces of Nature?" *AMBIO: A Journal of the Human Environment* 36, no. 8 (2007): 618.
4. Donna J. Haraway, "Situated Knowledges: The Science Question in Feminism and the Privilege of Partial Perspective," in *Simians, Cyborgs, Women: The Reinvention of Nature* (New York: Routledge, 1991), 188, 189.
5. Claire Colebrook, *Death of the Posthuman: Essays on Extinction, Vol. 1* (Ann Arbor, Mich.: Open Humanities Press, 2014), 22. This passing reference to Colebrook's work should by no means imply that it can be readily encapsulated. Indeed, I think her extensive, bold, and often disconcerting work on the concepts of extinction and the anthropocene make her the preeminent philosopher of these emerging fields of thought.
6. Andrew Revkin, "Confronting the Anthropocene," *New York Times*, May 11, 2011, http://dotearth.blogs.nytimes.com/2011/05/11/confronting-the -anthropocene/?_php=true&_type=blogs&_r=0.
7. Elizabeth Kolbert, "Age of Man," *National Geographic*, March 2011, http:// ngm.nationalgeographic.com/2011/03/age-of-man/kolbert-text.
8. "Anthropocene," *Encyclopedia of Earth*, September 3, 2013, http://www .eoearth.org/view/article/150125/.
9. Owen Gaffney, "State of the Art," *Anthropocene Journal*, October 1, 2013.
10. Ibid. The posting shows artwork and data visualizations by Félix Pharand-Deschênes, David Thomas Smith, Stephen Walter, Jason deCaires Taylor, Radhika Gupta, John Stockton, and NASA's Landsat program.
11. "Cartography of the Anthropocene," *Globaïa*, 2013, http://globaia.org/ portfolio/cartography-of-the-anthropocene/.
12. Jamie Lorimer makes a similar argument, calling for attention to what he terms "nonhuman mobilities": "Tracing networks maps the geographies of intersecting lines through which landscapes are to be reanimated

and by which their difference is threatened. . . . An attention to animals' geographies—thinking like an elephant, an insect, or even a molecule—can help attune to the diverse ways in which nonhuman life inhabits the novel ecosystems of an Anthropocene planet." See Lorimer, *Wildlife in the Anthropocene: Conservation after Nature* (Minneapolis: University of Minnesota Press, 2015), 177, 176.

13. Take, by contrast, Nicole Starosielski's multimedia project "Surfacing," a digital map of underwater cable systems in which "the user becomes the signal and traverses the network." See Starosielski, "Surfacing: A Digital Mapping of Submarine Systems," *Suboptic*, 2013, 3. "Surfacing" will be discussed below.

14. Betsy Wills, "'Anthropocene': Ariel Photography by David Thomas Smith," *Artstormer*, March 15, 2013, http://artstormer.com/2013/03/15/anthropocene-aerial-photography-by-david-thomas-smith/.

15. Ibid.

16. "There is no Space or Time / Only intensity, / And tame things / Have no immensity." Mina Loy, "There Is No Life or Death," in *The Lost Lunar Baedeker: Poems of Mina Loy*, ed. Roger L. Conover (New York: Farrar, Strauss and Giroux, 1997), 3.

17. Patricia Johanson, quoted in Xin Wu, *Patricia Johanson and the Re-invention of Public Environmental Art, 1958–2010* (Surrey, U.K.: Ashgate, 2013), 155.

18. One of the sections to follow will discuss remarkable exceptions that enmesh the human with the lithic.

19. Dipesh Chakrabarty, "The Climate of History," *Critical Inquiry* 35 (Winter 2009): 201.

20. Ibid., 206–7.

21. Jan Zalasiewicz, Mark Williams, Will Steffen, and Paul Crutzen, "The New World of the Anthropocene," *Environmental Science and Technology Viewpoint* 44, no. 7 (2010): 2229.

22. Chakrabarty, "Climate of History," 207.

23. Ibid., 220.

24. Ibid.

25. Elizabeth DeLoughrey, "Ordinary Futures: Interspecies Worlding in the Anthropocene," in *Global Ecologies and the Environmental Humanities: Postcolonial Approaches*, ed. Elizabeth DeLoughrey, Jill Didur, and Anthony Carrigan (New York: Routledge, 2015), 354.

26. Donna J. Haraway, *When Species Meet* (Minneapolis: University of Minnesota Press, 2007), 25.

27. W. E. B. Du Bois, *The Souls of Black Folk* (Mineola, N.Y.: Dover, 1994); Frantz Fanon, *Black Skin, White Masks* (New York: Grove Press, 1967); Homi Bhabha, *The Location of Culture* (New York: Routledge, 2004); Judith Butler, *Gender Trouble: Feminism and the Subversion of Identity* (New York: Routledge, 1999).

28. Rory Rowan, "Notes on Politics after the Anthropocene," *Progress in Human Geography* 38, no. 3 (2014): 449.
29. Ian Baucom, "The Human Shore: Postcolonial Studies in an Age of Natural Science," *History of the Present: A Journal of Critical History* 2, no. 1 (2012): 4. Many thanks to Sangeeta Ray for sending me this essay.
30. Jamie Lorimer, *Wildlife in the Anthropocene: Conservation after Nature* (Minneapolis: University of Minnesota Press, 2015), 3.
31. "Slow violence" is of course Rob Nixon's term from *Slow Violence and the Environmentalism of the Poor* (Cambridge, Mass.: Harvard University Press, 2011). Rob Nixon points out the painfully ironic timing of the "the grand species narrative of the Anthropocene," which is "gaining credence at a time when, in society after society, the idea of the human is breaking apart economically, as the distance between affluence and abandonment is increasing." He asks, "How can we counter the centripetal force of that dominant story with centrifugal stories that acknowledge immense disparities in human agency, impacts, and vulnerability?" See "The Great Acceleration and the Great Divergence: Vulnerability in the Anthropocene," *Profession*, March 19, 2014, https://profession.commons.mla.org/2014/03/19/the-great-acceleration-and-the-great-divergence-vulnerability-in-the-anthropocene/.
32. Sylvia Wynter and Katherine McKittrick, "Unparalleled Catastrophe for Our Species? Or to Give Humanness a Different Future: Conversations," in *Sylvia Wynter: On Being Human as Praxis*, ed. McKittrick (Durham, N.C.: Duke University Press, 2015), 24. I should note that this brief inclusion of Wynter's brilliant work does not address the many ways in which its original conceptions clash with other models of environmental and feminist science studies and material feminisms in this book, for example her idiosyncratic definition of the "biocentric" and her use of Darwin. Wynter critiques the idea that humans are "purely biological beings," arguing instead that humans are hybrid creatures of both "mythoi and bios" (34, 31). Critical posthumanist and animal studies scholars, including myself, would not agree with this human exceptionalist argument that denies nonhuman beings their own modes of culture. Wynter plainly states, for example, "As far as eusocial insects like bees are concerned, their roles are genetically *preprescribed* for them. Ours are not" (34). Such rigid distinctions are not only problematic for posthumanists, but also for new materialists, in that it is problematic to draw a sharp line between biological embodiment and culture, given their many intra-actions. Even genetics can no longer be seen as encapsulated within the "biological" since epigenetics means that social, political, and environmental factors alter bodies. See, for example, Shannon Sullivan, *The Physiology of Sexist and Racist Oppression* (Oxford, U.K.: Oxford University, 2015).
33. Ibid.

34. Alexander G. Weheliye, *Habeas Viscus: Racializing Assemblages, Biopolitics, and Black Feminist Theories of the Human* (Durham, N.C.: Duke University Press, 2014), 24.

35. Anna Lowenhaupt Tsing, *Friction: An Ethnography of Global Connection* (Princeton, N.J.: Princeton University Press, 2005), 8.

36. Chakrabarty, "Climate of History," 14.

37. Dipesh Chakrabarty, "Brute Force," *Eurozine*, October 7, 2010, http://www.eurozine.com/articles/2010-10-07-chakrabarty-en.html.

38. I am grateful to Karen Barad's critique of this sentence, during the October 2014 SLSA conference, and her suggestion that I consider the (hypothetical) graviton particle. The graviton has confused me, however, since, if the graviton does exist, it would be a particle but would have no mass. So, by saying "humans are not gravity," I intend to critique Chakrabarty's mystification of humans as an abstract force. Reading a bit of physics, including Barad's work, does not leave me with the sense that even gravity is not gravity in that it may not be an immaterial force, but instead that it remains a bit of a mystery. Barad states, "Constructing a quantum theory of gravity means understanding how to apply quantum theory to the general theory of relativity. This has proved exceedingly difficult." By contrast, it is not so difficult to demonstrate the many ways, from agriculture to automobiles to acidification, that humans have brought about the anthropocene. Karen Barad, *Meeting the Universe Halfway* (Durham, N.C.: Duke University Press, 2007), 350. There is a very good chance that my thin understanding of physics caused me to misunderstand Barad's critique.

39. Dipesh Chakrabarty "Postcolonial Studies and the Challenge of Climate Change," *New Literary History* 43, no. 1 (2012): 13.

40. Jessi Lehman and Sara Nelson in "After the Anthropocene: Politics and Geographic Inquiry for a New Epoch," *Progress in Human Geography* 38, no. 3 (2014): 444.

41. Derek Woods, "Scale Critique for the Anthropocene," *Minnesota Review* 83 (2014): 134,.

42. Ibid., 140.

43. Rosi Braidotti, *Transpositions* (Cambridge, U.K.: Polity, 2006), 278.

44. Starosielski, "Surfacing," 3.

45. Nicole Starosielski, Erik Loyer, and Shane Brennan, "Surfacing," n.d., http://www.surfacing.in/. See also Starosielski's book, *The Undersea Network* (Durham, N.C.: Duke University Press, 2015).

46. Nicole Starosielski, *The Undersea Network* (Durham, N.C.: Duke University Press, 2015), 2.

47. Ibid., 2–3.

48. Stacy Alaimo, *Bodily Natures: Science, Environment, and the Material Self* (Bloomington: Indiana University Press, 2010).

49. See ibid., 119–25; and Rhonda Zwillinger, *The Dispossessed: Living with Multiple Chemical Sensitivities* (Paulden, Ariz.: Dispossessed Project, 1998).

50. Claire Colebrook, "Not Symbiosis, Not Now: Why Anthropogenic Climate Change Is Not Really Human," *Oxford Literary Review* 34, no. 2 (2012): 198–99.

51. Ibid., 193.

52. Ibid.

53. Jeffrey J. Cohen, *Stone: An Ecology of the Inhuman* (Minneapolis: University of Minnesota Press, 2015), 6, 62.

54. Elizabeth Ellsworth and Jamie Kruse, eds., *Making the Geologic Now: Responses to Material Conditions of Contemporary Life* (Brooklyn, N.Y.: Punctum Books, 2013), 152.

55. Ibid., 25.

56. Ilana Halperin, "Autobiographical Trace Fossils," in *Making the Geologic Now*, ed. Ellsworth and Kruse, 156.

57. Ibid.

58. Kathryn Yusoff, "Geologic Life: Prehistory, Climate, Futures in the Anthropocene," *Environment and Planning D: Society and Space* 31, no. 5 (2013): 780.

59. Stephanie LeMenager, *Living Oil: Petroleum Culture in the American Century* (New York: Oxford University Press, 2014), 6.

60. For another figuration of the anthropocene ocean, see DeLoughrey's "Ordinary Futures," which reads New Zealand Maori author Keri Hulme's speculative fiction by way of deep seabed mining, proposing that "we might read Hulme's oceanic imaginary in line with a cultural politics that destabilizes the state claims of the Foreshore and Seabed Act (and the Marine and Coastal Area Bill), a way of narratively imagining a relationship to the oceanic through ordinary modes of merger and submersion—an adaptive, interspecies hermeneutics for the rising tides of the anthropocene" (367).

61. See Stacy Alaimo, "New Materialisms, Old Humanisms; or, Following the Submersible," *NORA: Nordic Journal of Feminist and Gender Research* 19, no. 4 (2011): 280–84.

62. Take, for example, James Cameron's *Aliens of the Deep* (2005), a documentary about deep-sea exploration that repeatedly supplants the seas with the planets. The deep seas are cast as the perfect practice arena for space explorers, marine biology is said to be a good starting point for astrobiology, and the samples from the ocean are the "next best thing" for the planetary scientist to examine. The ethereal trumps the aqueous, the transcendent transcends the immanent. Marine biologist Dijanna Figuero's compelling and informative discussion of symbiosis in riftia (giant tube worms), for example, is followed by a cut to Cameron telling a scientist, "The real question is, can you imagine a colony of these on [Jupiter's moon] Europa?" Stacy Alaimo, "Dispersing Disaster: The Deepwater Horizon, Ocean Conservation, and the Immateriality of Aliens," in *Disasters, Environmentalism, and Knowledge*, ed. Sylvia Mayer and Christof Mauch (Heidelberg, Ger.: Universitätsverlag, 2012), 175–92.

63. Lesley Evans Ogden, "Marine Life on Acid," *BioScience* 63, no. 5 (2013): 322.
64. Ibid.
65. Ibid., 328.
66. Ibid.
67. Ibid., 323.
68. James C. Orr et al., "Anthropogenic Ocean Acidification over the Twenty-First Century and Its Importance to Calcifying Organisms," *Nature*, September 29, 2005, 685.
69. Ogden, "Marine Life," 323.
70. Jason Bidel, "Our Climate Change, Ourselves," *On Earth*, May 6, 2014, http://www.onearth.org/articles/2014/05/national-climate-assessment; National Climate Assessment report, U.S. Global Change Research, 2014, http://nca2014.globalchange.gov/downloads; Scott K. Johnson, "Sea Butterflies Already Feeling the Sting of Ocean Acidification?" *Ars Technica*, November 27, 2013, http://arstechnica.com/science/2012/11/sea-butterflies-already-feeling-the-sting-of-ocean-acidification/.
71. NOAA website, "What Is Ocean Acidification?" n.d., http://www.pmel.noaa.gov/co2/story/What+is+Ocean+Acidification%3F; Richard A. Kerr, "Ocean Acidification: Unprecedented, Unsettling," 2010, on the Ukrainian Science Club website http://nauka.in.ua/en, originally published in *Science*, June 18, 2010, 1500–1501.
72. Julia Whitty, "Snails Are Dissolving in Acidic Ocean Waters," *Mother Jones*, November 2012, http://www.motherjones.com/blue-marble/2012/11/first-evidence-marine-snails-dissolving-acidic-waters-antarctica. Tim Senden of the Department of Applied Maths at the Research School of Physics and Engineering, Australian National University, produced this video, which is available on YouTube at https://www.youtube.com/watch?v=48qrlTFqelc. Information about complex technologies and procedures of the X-Ray CT Lab is available at http://www.anu.edu.au/CSEM/machines/CTlab.htm.
73. Melissa Smith, "Climate Change as Art," *Australian Antarctic Magazine* 25 (December 2013), http://www.antarctica.gov.au/about-us/publications/australian-antarctic-magazine/2011-2015/issue-25-december-2013/art/climate-change-as-art.
74. Jellyfish and other gelatinous creatures, for example, have been portrayed as "art" in museum exhibits, coffee table books, videos for relaxation, and scientific and popular websites. See Stacy Alaimo, "Jellyfish Science, Jellyfish Aesthetics: Posthuman Reconfigurations of the Sensible," in *Thinking with Water*, ed. Janine MacLeod, Cecilia Chen, and Astrida Neimanis (Kingston, Ont.: McGill-Queen's University Press, 2013), 139–64.
75. Wikipedia, "Lysergic acid diethylamide," n.d., http://en.wikipedia.org/wiki/Lysergic_acid_diethylamide.

76. Richard M. Doyle, *Darwin's Pharmacy: Sex, Plants, and the Evolution of the Noösphere* (Seattle: University of Washington Press, 2011), 20.
77. Ibid., 21.
78. DeLoughrey, "Ordinary Futures," 365.
79. Rosi Braidotti, *The Posthuman* (Cambridge, U.K.: Polity, 2013), 134.
80. Ibid., 136.

Conclusion

1. Jorie Graham, "Sea Change," in *Sea Change* (New York: HarperCollins, 2008), 3.
2. See Stacy Alaimo, *Undomesticated Ground: Recasting Nature as Feminist Space* (Ithaca, N.Y.: Cornell University Press, 2000), 63–70.
3. Ibid.
4. Given that sustainability has become such a pervasive paradigm within the United States, it is ironic that the term itself originated as part of "sustainable development" movements. Catriona Sandilands notes that even though the word "development" has largely disappeared it is "very much still the ghost animating the rhetoric" (personal communication).
5. John P. O'Grady, "How Sustainable Is the Idea of Sustainability?" *ISLE: Interdisciplinary Studies in Literature and Environment* 10, no. 1 (2003): 3.
6. Slavoj Žižek, *Living in the End Times* (London: Verso, 2011), 328.
7. U.S. EPA website, "Environmental Management Systems," n.d., http://www.epa.gov/EMS/.
8. See Samuel P. Hays, *Conservation and the Gospel of Efficiency: The Progressive Conservation Movement, 1890–1920* (Pittsburgh: University of Pittsburgh Press, 1999).
9. Stephanie LeMenager and Stephanie Foote, "The Sustainable Humanities," *PMLA* 127, no. 3 (2012): 572–73.
10. Daniel J. Philippon, "Sustainability and the Humanities: An Extensive Pleasure," *American Literary History* 24, no. 1 (2012): 164–66.
11. Sally Kitch, Joni Adamson, and members of the Faculty Working Group on Humanities and Sustainability, *Contributions of the Humanities to Issues of Sustainability* (Tempe: Arizona State University, 2008).
12. Gert Goeminne, "Once upon a Time I Was a Nuclear Physicist: What the Politics of Sustainability Can Learn from the Nuclear Laboratory," *Perspectives on Science* 19, no. 1 (2011): 20.
13. LeMenager and Foote, "Sustainable Humanities," 574.
14. See Phil Brown for a definition of "popular epidemiology" and Giovanna Di Chiro for a definition of "ordinary experts." Brown, "When the Public Knows Better: Popular Epidemiology," *Environment* 35, no. 8 (1993): 17–41; Di Chiro, "Local Actions, Global Visions: Remaking Environmental Expertise," *Frontiers* 18, no. 2 (1997): 203–31.

15. See Stacy Alaimo, "Deviant Agents," in *Bodily Natures: Science, Environment, and the Material Self* (Bloomington: Indiana University Press, 2010); and the film *Tree-Sit: The Art of Resistance*, dir. James Ficklin (Earth Films, 2005).

16. Rosi Braidotti, *Transpositions* (Malden, Mass.: Polity, 2006), 137.

17. Ibid., 278.

18. Žižek, *Living in the End Times*, 424.

19. See Alaimo, "Material Memoirs," in *Bodily Natures*, 85–112.

20. Philippon, "Sustainability and the Humanities," 175.

21. Julian Agyeman, "Toward a 'Just' Sustainability?" *Continuum: Journal of Media and Cultural Studies* 22, no. 6 (2008): 752.

22. Rob Nixon, *Slow Violence and the Environmentalism of the Poor* (Cambridge, Mass.: Harvard University Press, 2011).

23. United Nations World Commission on Environment and Development, *Our Common Future* (Oxford, U.K.: Oxford University, 1987), 43.

24. Bill McKibben, *The End of Nature* (New York: Random House, 1989).

25. Žižek, *Living in the End Times*, 481.

26. Denise L. Breitburg et al., "Ecosystem Engineers in the Pelagic Realm: Alteration of Habitat by Species Ranging from Microbes to Jellyfish," *Integrative and Comparative Biology* 50, no. 2 (2010): 197.

27. Ibid., 198.

28. Karen Barad, *Meeting the Universe Halfway* (Durham, N.C.: Duke University Press, 2007), 393.

29. Levi R. Bryant, *Onto-cartography: An Ontology of Machines and Media* (Edinburgh: Edinburgh University Press, 2014), 215.

30. Similar to "equality" would be Levi Bryant's use of the phrase "democracy of objects," which he clearly delineates: "The democracy of objects is not a political thesis to the effect that all objects ought to be treated equally or that all objects ought to participate in human affairs. The democracy of objects is the ontological thesis that all objects, as Ian Bogost has so nicely put it, equally exist while they do not exist equally. . . . In short, no object such as the subject or culture is the ground of all others." See Bryant, *The Democracy of Objects* (Ann Arbor, Mich.: Open Humanities Press, 2011), 19.

31. Ian Bogost, *Alien Phenomenology; or, What It's Like to Be a Thing* (Minneapolis: University of Minnesota Press, 2012), 6.

32. Ibid., 10.

33. Ibid., 7.

34. Ibid., 3.

35. Barad, *Meeting the Universe Halfway*, 337.

36. Bryant, *Onto-cartography*, 63.

37. Ibid., 37.

38. Ibid., 71.

39. Bogost, *Alien Phenomenology*, 9–10.
40. Ibid., 10.
41. Jeffrey Jerome Cohen offers a more generous, and allied, reading of object-oriented philosophy as a mode of (speculative) realism: "Yet its realism is *weird*, meaning that this world is not reducible to common sense, the evidence of the mind, or other modes of imposing human order. It thereby shares expressive affinities with speculative fiction—science fiction, horror, and fantasy. Medieval versions of these wonder-inducing genres include romance, lays, and lapidaries, providing those who study the Middle Ages rich entrance into this critical conversation." See *Stone: An Ecology of the Inhuman* (Minneapolis: University of Minnesota Press, 2015), 45.
42. Jacques Derrida, *The Animal That Therefore I Am* (New York: Fordham University Press, 2008), 32.
43. Andrew Pickering, *The Mangle of Practice: Time, Agency, and Science* (Chicago: University of Chicago Press, 1995).
44. Kim TallBear, "An Indigenous Reflection on Working beyond the Human / Not Human," *GLQ* 21, nos. 2–3 (2015): 233.
45. Ibid.
46. Timothy Morton, *Hyperobjects: Philosophy and Ecology after the End of the World* (Minneapolis: University of Minnesota Press, 2013), 18.
47. Ibid., 18. I do not know which "ecophenomenologists" Morton is targeting here, since none is listed in his endnotes. I do not characterize my own work as "ecophenomenology," but rather as part of new materialism, material feminisms, ecomaterialisms, and science studies. See the introduction of this book for a brief discussion of the limits of phenomenology.
48. Ibid., 27–29.
49. Nancy Tuana, "Viscous Porosity: Witnessing Katrina," in *Material Feminisms*, ed. Stacy Alaimo and Susan J. Hekman (Bloomington: Indiana University Press, 2008), 188. Morton does not cite Nancy Tuana or Karen Barad in *Hyperobjects*. The chapter after "Viscosity" is titled "Nonlocality," and it includes descriptions that parallel my notion of trans-corporeality and Tuana's "viscous porosity": "Likewise, endocrine disruptors penetrate my body through my skin, my lungs, and my food. The disruptors in pesticides such as Roundup, a cousin of Agent Orange (also made by Monsanto), often dioxins of some kind, start cascading reactions in my body, interfering with the production and circulation of hormones." See Morton, *Hyperobjects*, 38. This description could be extended to include the effects of Roundup on the workers who manufacture it and the nonhuman creatures that encounter it, rather than reducing the scene to an isolated human and an isolated hyperobject.
50. Tuana, "Viscous Porosity," 202.
51. Ibid., 203.

52. Eli Clare, "Meditations on Natural Worlds, Disabled Bodies, and a Politics of Cure," in *Material Ecocriticism*, ed. Serenella Iovino and Serpil Oppermann (Bloomington: Indiana University Press, 2014), 205.

53. For a study of the relation between disability and environmentalism see Sarah Jaquette Ray, *The Ecological Other: Environmental Exclusion in American Culture* (Tempe: University of Arizona Press, 2013).

54. Ibid., 218.

55. Beatriz Preciado, *Testo Junkie: Sex, Drugs, and Biopolitics in the Pharmacopornographic Era* (New York: Feminist Press, 2013), 114, 79.

56. Cary Wolfe, *What Is Posthumanism?* (Minneapolis: University of Minnesota Press, 2010), xvi.

57. Eva Hayward, "More Lessons from a Starfish: Prefixial Flesh and Transspeciated Selves," *WSQ: Women's Studies Quarterly* 36, nos. 3–4 (2008): 81–82.

58. Wolfe, *What Is Posthumanism?* xxix.

59. Ibid., 65, 82.

60. Susan J. Hekman, *The Material of Knowledge: Feminist Disclosures* (Bloomington: Indiana University Press, 2010), 93.

61. Hayward, "More Lessons from a Starfish," 69, 70.

62. Morton, *Hyperobjects*, 17.

63. Bogost, *Alien Phenomenology*, 124, 134.

64. See Sara Ahmed's *The Promise of Happiness* (Durham, N.C.: Duke University Press, 2010). Bogost includes a story about an image of a woman in a Playboy Bunny outfit, which betrays the fear that females have no sense of humor or intellectual sophistication. See *Alien Phenomenology*, 93–99. Feminist and queer theory and disability studies have themselves developed playful, pro-sex understandings of bodies, objects, pleasures, and desires that would contest simple notions of sexual objectification. Given this exciting body of work—not to mention the lively sexual (sub)cultures with which the theories are allied—it's hard to imagine that we ("women or girls or sexiness") need OOO to give us an "ontological place alongside chipmunks, lighthouses, and galoshes" (99).

65. Bogost, *Alien Phenomenology*, 24.

66. To find the ecosexuals, see chapter 3.

67. Pam Longobardi, "Drifters Project Statement," 2016, http://driftersproject.net/about/.

Index

◇◇

Stacy Alaimo is professor of English and Distinguished Teaching Professor at the University of Texas at Arlington. She is author of *Undomesticated Ground: Recasting Nature as Feminist Space* and *Bodily Natures: Science, Environment, and the Material Self*; coeditor of *Material Feminisms*; and editor of *Matter*.